Ways of Renewal
a Guidebook for Women

Krisztina Samu, LAc

© 2018 Lulu   Krisztina Samu, LAc - ProjectAcuhope - www.projectacuhope.com

ISBN:978-0-578-21481-8

Cover Image © Colby D. Kuschatka

Thank you

To Frank for encouragement, love and financial support.

To the South Brunswick Police Department for taking my complaint
and doing nothing even if they believed me, leading me to question everything.

To Victor for offering protection and guidance.

To Sue who by the power of her presence confirmed to me the
efficacy of acupuncture in the treatment of rape trauma.

To John P. Urich, LAc, my first acupuncturist who administered treatments and facilitated my healing.

To J. R. Worsley, Lonny Jarret and other knowledge keepers who carry the ancient wisdom of
Oriental Medicine into the modern world.

To Norma Rawlins, LAc who taught me about Ghost Points.

To Dr. Chieko Maekawa who trained me in the Seitai Shinpo Japanese Acupuncture Lineage.

To Kevin Quirk from yourbookghostwriter.com for his hard work on this manuscript.

And to my parents for their tireless patience.

Table of Contents

Note From the Author

They say that time heals all wounds. That said, not everything people say is necessarily true.

When I was a 24 year old woman, and I suffered a sexual assult. In the aftermath, I began to ask myself... "Something broke.... what is it? Something very harmful has been left inside of me.... what is it? If I can figure out what it is, I can figure out how fix it." That is what led me on a long and interesting path of victim to survivor to acupuncturist to now author.

Regardless of the precise circumstances of the sexual trauma that you experienced, healing is possible. I trust that the information contained in this book will guide you to a deeper understanding of what happened in your body and how to heal it. My goal is to make your healing path less of an excruciatingly long winding road with many forks and dead ends. I wish for you a much shorter path to wholeness, similar to what a person travels when they break an arm and simply go to the emergency room for the appropriate treatment.

Krisztina Samu, LAc
Silver Certificate holder, Seitai Shinpo Acupuncture
Holder of Light in the Tunnel of Darkness

© Can Stock Photo / tanais

Preface

There are multiple factors that influence our healing journey of recovery from sexual trauma. Each survivor's history of trauma is unique and the healing journey will be too!

- The age that the violation happened.
- The number of times the violation happened.
- The relationship of the perpetrator(s) to the victim.
- The intensity of the violation.
- The survivor's inner resources (inner strength, resilience and commitment to heal).
- The survivor's outer resources (peer or family support, therapeutic support, financial resources).
- The life circumstances of the survivor with respect to family, friends, home and work when she embarks on the healing journey.

There are many influencing factors which may complicate the picture. Some survivors may have grown up in a relatively stable and supportive family and are struggling with a single incident of sexual assault by a perpetrator they did not know well. Other survivors may have been repeatedly raped from childhood by a close family member. They may have grown up in a dysfunctional family and have intermittent struggles with substance abuse, and lack financial resources or a safe place to live. Everyone's situation is different.

If your particular situation is complex, do not be disheartened. Everything does not have to be addressed at once. It is possible to heal one set of issues at a time and attend to the other issues when circumstances allow. Each step towards bettering one's situation in one area has the ripple effect of positively affecting the other areas.

The focus of this book is on body-centered healing techniques which will have a positive effect on healing the body, emotions and spirit. It is beyond the scope of this book to address other influencing factors such as healing of family or peer relationships, overcoming dependence on drugs or alcohol, resolving matters of housing or employment and many other matters that may complicate recovery. To balance those life factors requires efforts that I do not discuss within this guidebook.

Your intuition will guide you to the information that is most helpful for you. Trust your inner guidance to be drawn to the information that is most relevant to you based on where you find yourself at in the present. Just keep reading and see where you find your "aha's."

Each case of sexual assault is unique and, as such, each healing path will be unique. This book does not attempt to outline a single formula which will help all cases equally. Rather, this book presents a menu of choices based on common symptoms experienced by survivors.

Body-centered healing techniques are an excellent way of clearing energy blocks in the body of a traumatized person, so the survivor's emotions harmonize and her true spirit can once again take residence in the body and feel at home.

The healing techniques in this book will be equally useful for women who have experienced sexual depletion, which is a sort of emptiness due to too much loveless sex (called fornication in religious texts) that leaves a person feeling empty inside. Similarly, male survivors may find certain healing modalities described in this book to be useful, although the book does reference a few therapies specific to the uterus which, naturally, are for females only.

Introduction

Ages ago, people who broke a bone typically lost full and proper use of their arm or leg until bonesetters figured out how to set the bone and then immobilize the limb so it could heal properly. Similarly, advances in reconstructive surgery turned debilitating disfigurement from injuries or birth defects that had long been considered "untreatable" into something that no longer came with a sentence of doom. Today, because of greater understanding of how trauma presents in the body, and an intersection between ancient therapies and modern science, we have enough information to heal rape trauma, which for most of human history was not regarded or acknowledged as something that needed to be treated. As a result of this thinking, many women suffered in our past. As such, today, the horizons are opening so that we are able to transform the lives and suffering of many sexual trauma survivors and lead them into a brighter future.

Living as a survivor of rape trauma, or in the legacy of any sexual assault or abuse, can feel like coping with the aftermath of a very private war. Most people around you can't quite understand the nature of your suffering. They may not want to hear about it, and you may not feel like talking about it anyway. You do your best to deflect any harsh and judgmental questions that may be directed at you: Did you in any way encourage what happened? Did you resist hard enough? Isn't it time to let go, move on, get over it? If so, why is this so difficult?

Talking to a confidante, seeing a therapist or joining a therapy group of survivors may provide a degree of relief and understanding. Yet, no matter what you may do to reach out for help, or how long it's been since you were sexually assaulted, you may find yourself adrift in cycling emotions from the past or have internalized deep negative emotions that don't seem to transform and resolve, or you may just feel numb, stuck or lost. You know you're not yourself, that you've lost your unique spark, but you can't seem to find the answer of how to change your circumstances. You wonder if this suffering is ever going to end, if you'll ever return to a life that feels whole and fulfilling.

Perhaps you can relate to parts of that description of life as a survivor. Or maybe your pain is more subtle, quieter, harder to articulate. You sense that something troublesome is going on inside you, but you can't name it. You wake up some mornings with a familiar dread of facing one more day when you can't trust people and you can't trust the world, and you just want to pull the covers over your head and not move. And even if you have tried therapy or medication, that deep sense of sadness, fear or rage doesn't really seem to be getting better.

I understand that kind of quiet suffering, those deep levels of frustration, that creeping sense of hopelessness. Many years ago, as a young woman in my twenties, I was raped, and for quite a long time afterward I suffered. But I knew, I absolutely knew that there must be a way to heal. Then, a profound and effective path of healing emerged in front of me and I decided then that just as we now know what to do for someone who breaks a bone, it must be possible to find effective body-centered treatment protocols for those who have experienced sexual trauma.

Today, my life is driven by an unwavering commitment to share my twenty years of reflection and research to help survivors of sexual trauma. I want to help you come to believe that no matter how stuck or hopeless you may feel today, a completely different and positive future is possible for you. I'm going to do everything I can to help you better understand what really happens to our bodies on an energetic level after rape. I will offer you possibilities of how you might choose to pursue your own healing, introducing you to approaches and ideas that may alleviate your suffering and point you toward health and wholeness. I passionately believe that there's a lot more out there in terms of effective resources for rape trauma than most people know about, and you should have full access to that information. The map of opportunities for you to claim healing should be expanded wider and wider. You deserve nothing less!

Through this book, I also hope to become a part of a change in our culture's attitudes, understanding and beliefs about the trauma women suffer after being violated, and the most effective ways to support and assist survivors. I honestly believe we can reach a place where we regard rape as a "treatable injury," rather than a permanent wound that negatively drives our destiny.

If you have only recently endured a sexual assault, my hope is that this book will serve as a springboard for you to discover your path to recovery now, so that you may avoid years of suffering. Just as a broken bone that is not properly set and immobilized may result in a long and incomplete healing process, trauma from sexual assault that is not addressed and treated in the most effective way, ideally soon after the assault, can leave you "broken" for far too long. However, even if your sexual trauma occurred months or even years ago, deep healing is still very much available to you. My wish for you is that reading this book will mark the beginning of the end of your pain.

I also welcome the loved ones of rape trauma survivors, those who offer their support and compassion for the healing journey. By learning more about the many treatment opportunities that the person you love may reach out for, you will be in a stronger place to offer encouragement and hope.

If you are a professional caregiver, no matter what your belief system or therapeutic approach may be, I offer my ideas, experience and perspective to you as well. I hope that what I share in these pages may in some way benefit you in your own work with survivors of sexual assault.

In the chapters ahead, we will be looking closely at the impact of rape trauma on your body, mind and soul. We will explore the limits of mainstream talk therapy in treating rape trauma, and why even extensive talk therapy may leave a survivor feeling that she is not fully healed. After all, the trauma lives on in the body. It is not isolated to the psyche. We will discuss how acupuncture and Oriental Medicine are especially effective in treating the invisible injury of rape trauma, and also delve into further discussion to explore treatment modalities that are particularly effective.

In the section on Oriental Medicine, I will be sharing my ideas and experience from the lens of a licensed acupuncturist who assists women dealing with a wide spectrum of conditions, including symptoms originating from sexual assault. But my connection to rape trauma and acupuncture goes deeper than that. Earlier, I mentioned that I was raped many years ago and grappled with the effects for several difficult years until I had a personal breakthrough.

At that particular point of my life, I was attending a support group for people wishing to better their interpersonal relationships. Over the two years that I attended, it turned out that sexual assault was a running theme among many of the group's participants. It was a common story and all of the individuals who disclosed this history still showed significant pain when they discussed the traumatic event as if, in some way, the trauma was still affecting their lives. My attention was particularly drawn in by the story of a woman in this group who had been raped not once but multiple times in her life. She also had endured a dysfunctional and abusive childhood. And yet, this woman exhibited greater emotional balance and inner strength than anyone in the group. When she spoke, she radiated an angelic glow that seemed completely out of place given her history. I had to ask her: what did she do to dramatically turn her life around? When she told me that she had seen an acupuncturist once a week, as a complement to working with a traditional therapist, I immediately asked for his name and made an appointment. And from my first acupuncture session, I knew that I had found my way home, my path to the healing of body, mind and spirit. I discovered first-hand that rape trauma resides in specific places in the body—strongly affecting the energy pathways called meridians in acupuncture—and that it really can be approached as a treatable injury. I was so impressed by the profound changes that acupuncture helped facilitate in my life, I soon made the decision to train to become an acupuncturist myself. Many years later, I learned of three other female acupuncturists, one in Florida, one in New Mexico and one in California who also had discovered acupuncture because they were seeking healing from sexual assault. They too had used this ancient modality to heal themselves and then became acupuncturists who were passionate in helping others. Our shared experiences confirmed my belief in acupuncture as a powerful tool in healing sexual assault.

Acupuncture is not the only pathway toward overcoming trauma, regaining emotional balance and experiencing multi-dimensional healing for those who have suffered sexual assault. What I can say is that I know that it can, without a doubt, profoundly help. I'll share with you why I believe that's true by providing a clear view of how Oriental Medicine successfully treats rape trauma.

I will explain some of the relevant inner workings of acupuncture in a way that you can understand, whether you have some experience or familiarity with body-oriented therapeutic work or no experience at all.

I will also devote a separate chapter to introducing you to many other body-oriented therapies that can become additional options for you in pursuit of your own healing. Once you start looking into these possibilities, I trust that your inner guidance will tell you what's right for you.

Finally, the book will dive into the challenging terrain of how rape trauma survivors may begin to reclaim their true essence, and live not in fear or hiding, but with their full power as women. Of course, there are significant gender and cultural obstacles to deal with in that process covered later in this book. For now, with God's help, I invite all of us to imagine and begin to create a better way of life—and a better world.

So let's begin!

Chapter 1
The Invisible Injury

Physical injuries are usually pretty easy to see. If we fall and break our ankle, the swelling, bruising, discoloration and deformity are right there in front of our eyes. What we can see from the outside or inside, in the event of surgery, is called gross anatomy. We know just what you're dealing with, and a well-trained doctor or surgeon may be called upon to perform the needed treatment. The path that will lead toward healing is marked by very clear directions.

Treating the injury of rape trauma is very different. It requires an understanding of a very different kind of anatomy, relating to the movement of subtle energies in the body. This anatomy, sometimes called esoteric anatomy, other times difficult to name, is nonetheless very real. Energy pathways have been mapped and named thousands of years ago in ancient, Eastern medical systems. These energy pathways include the meridians and acupuncture points of Oriental Medicine and the chakras and nadis of Ayurvedic medicine. Just as a car has a body and an electrical system, we have a body and an electrical system—a bio-electric system if you will. The trauma of rape resides primarily in the bio-electrical system of the body, as I will explain later in this book.

If you did suffer outward physical injuries during the sexual assault, they are likely to be visible and can be treated via the conventional medical route. For most survivors, however, whether they had any outward physical injuries or not, the most severe and potentially debilitating injuries are not so easy to detect. You can't immediately see the wound, at least physically. It's not an outward injury, it's an inward injury of normal function of the body's subtle energy systems. In other words, it's invisible. That means that before you can actively and effectively treat this injury, you need to

The bio-electric energy pathways of the body's acupuncture meridian system.

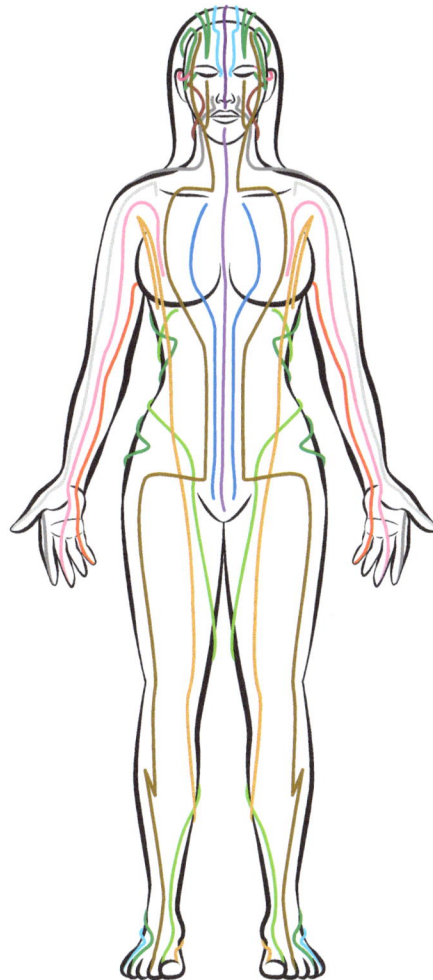

© Can Stock Photo / PeterHermesFurian

know where to find it, how to identify it, how it is operating inside you and what kinds of methods and approaches will work best in addressing your multi-leveled wound.

As survivors, we may already be familiar with the more visible symptoms of sexual trauma, either because we can identify them ourselves or because they have been pointed out to us by a therapist or people who are close to us. We may find ourselves depressed or crying frequently. Perhaps we lash out in anger at others. We might suffer from headaches, insomnia and/or nightmares, tightness in our chest or gastrointestinal disturbance. Differences in our daily routines may show up as a lack of appetite or chronic over-eating, or a sudden and uncharacteristic usage of alcohol or drugs. People who know us may react to how fearful or anxious we seem, and we may be withdrawing from others out of a belief that people and the world are no longer safe.

These are just some of the many symptoms that rape trauma survivors and healthcare professionals who work with them have described. It's important to recognize these kinds of symptoms when they surface, but it's even more important to understand that just naming them does not tell the whole story. We need to become better detectives in fleshing out where such symptoms really emanate from, what they represent and what they are telling us about what we actually need to get better. At the same time, we need to open up our viewfinders and become aware of the many other aspects and dimensions of this injury that do not come with psychological terms such as anxiety, depression, grief, shame or despair. To do so, it may be useful to expand our language and incorporate new terms or descriptions.

Remember, we're dealing with a mostly invisible injury. That's primarily because the injury of rape trauma is not showing up on the outside of our bodies like that broken ankle. Instead, it's causing suffering because of what's going on inside our bodies and minds.

Suffering sexual assault is like having your house broken into.

Many survivors describe rape as a personal invasion. Something that is ours, something that feels familiar and private and that comes with a sense of comfort and safety, truly has been invaded. It leaves us in shock and anger. How could someone have done this? What has been invaded or intruded upon, of course, is our body. But the traumatic experience goes even deeper than that. In psychology, "house" is often regarded as a metaphor for the self. In other words, what was broken into was really our whole being. Some experience it as a forcible entry into the soul.

We can take this analogy further by paying attention to the attitudes you adopted after you were violated. Do you find yourself turning to the need for "increased security?" In the aftermath of sexual assault, that may take the form of a kind of hypervigilance: "No one is going to break in here again!" Or you beef up your "armoring" in dealing with men and almost everyone in your life. Maybe you isolate yourself behind the beefed-up security and literally become shut down, as if your "house" has been barricaded at every door and window to prevent any further entry.

Some survivors respond differently to having their "house" (the body) broken into. They decide to adopt an "open door" policy. That can translate into sexually promiscuous behavior, where a survivor places herself in situation after situation where she enters into sexual activity with men she hardly knows. If your behavior has turned in this direction, you may be operating under a belief that the best way to avoid another "break-in" is to leave your door open so anyone can come in without needing to barge their way through. You may be clinging to a misguided notion that the sexual act that follows will not come with the same emotional force and trauma as the initial sexual assault. Another scenario is that an experience that left you feeling broken or defiled makes you feel that you only deserve subpar situations. You may feel that once knocked down you can't get up, or perhaps that your destiny now is disgrace. None of this has to be true! None of this has to be true!

Of course, neither of these strategies really gets to the heart of the matter. If your real house had been broken into, you would need to first clean things up and repair what's damaged. In your life after sexual assault, you need to energetically clean your body and heal what's broken.

Let's look more closely at this sense of being broken as it relates to rape trauma. We're all familiar with the term "broken heart." If someone complaining of a broken heart went to a surgeon who opened up that person's chest cavity, would he or she see the person's heart as being physically broken? Obviously not. So what is it that's actually broken? In acupuncture and other body-oriented therapies where practitioners work with the human energy field, we would refer to what is broken as the etheric field of the heart. In other words, the heart has a gross anatomy and a subtle anatomy; the gross anatomy is the muscle of the heart and the subtle anatomy is the etheric field that is vulnerable to emotional disturbance, like the pain suffered when someone has rejected or hurt you to the point where your heart feels as if it has been torn apart. The positive manifestation of this concept comes when we talk about "opening our heart" to someone we love. Physically, our heart didn't change size or shape or have a different part revealed or exposed. It's the etheric field of the heart, or the force field that acts as a boundary to the deeper emotional world of the heart,

that becomes more dense in a closed heart, or more subtle and permeable in an open heart.

If discussion of the human energy field is new to you, I like this definition by medical intuitive, Carolyn Myss: *"Your physical body is surrounded by an energy field that extends as far out as your outstretched arms and the full length of your body. It is both an information center and a highly sensitive perceptual system....It surrounds us and carries with us the emotional energy created by our internal and external experiences—both positive and negative."*

Barbara Brennan, a renowned healer, author and scientist, helps us understand the sense of being broken. She describes an energetic pathway called the Hara Line, a central axis that traverses our crown chakra at the highest point of our head through the root chakra in the perineum, extending up and down, beyond these two chakras. If the Hara Line is broken, it can cause major damage to our whole being. A broken Hara Line will make us feel broken to the core, and it also makes it difficult to regenerate sexual energy.

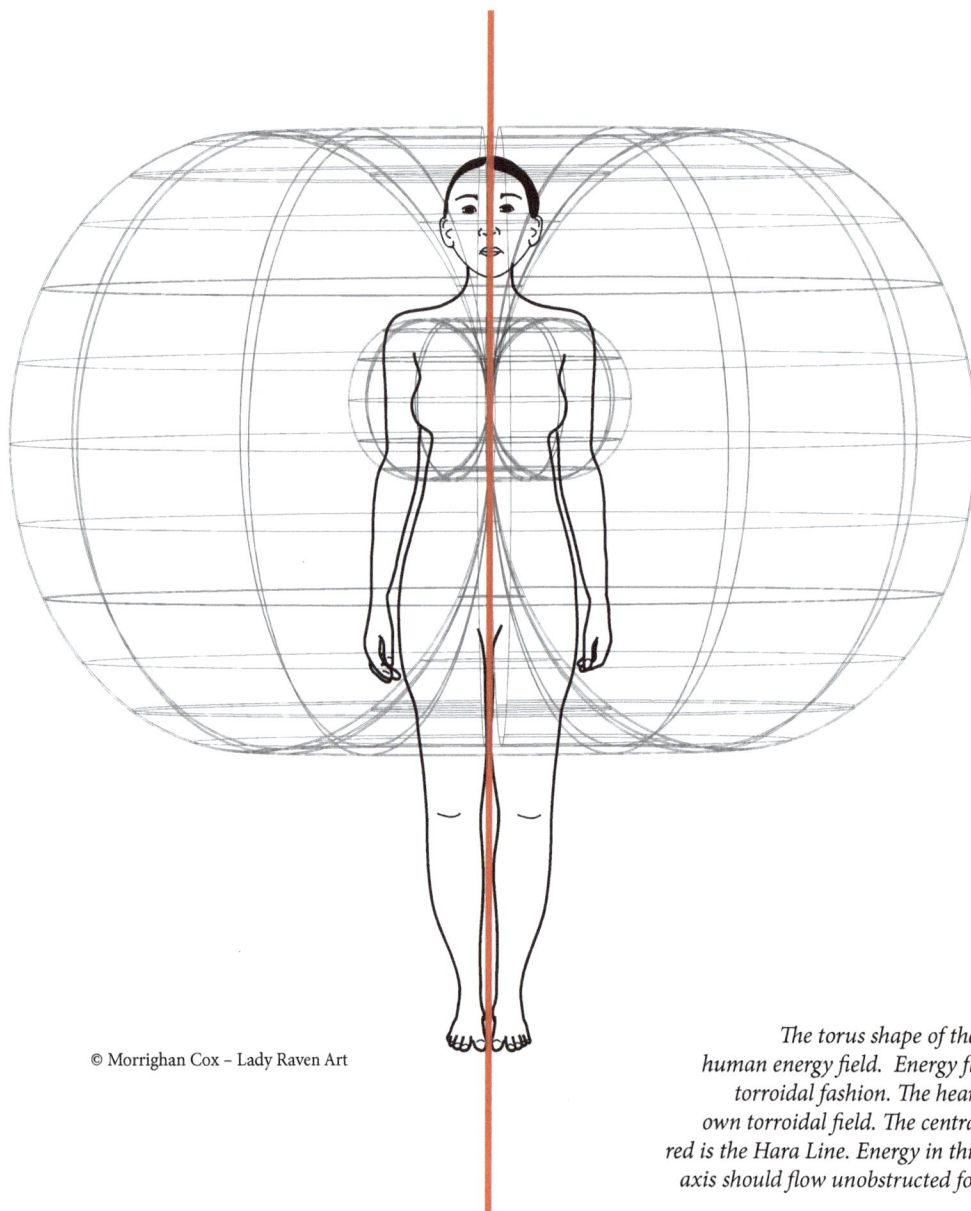

© Morrighan Cox – Lady Raven Art

The torus shape of the healthy human energy field. Energy flows in a torroidal fashion. The heart has it's own torroidal field. The central axis in red is the Hara Line. Energy in this central axis should flow unobstructed for health.

Have a look at this image of a leaf. The leaf has many of what we might call veins and capillaries that carry water and nutrients. The central vein, if you will, is the equivalent of the Hara Line. However, in a leaf, the veins and capillaries are visible. Our energetic pathways are not. In leaves these pathways carry water and nutrients, but in humans these similar systems carry Qi (pronounced chee), which is bioelectric energy or spiritual energy.

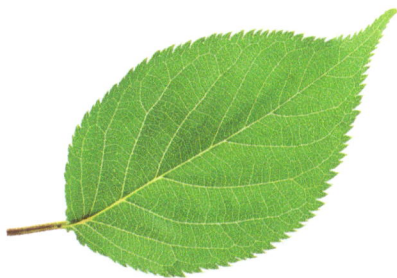

©IGORKO - Can Stock Photo Inc.

Looking at this concept through a different lens, consider how in outdated and barbaric traditions of horse training they talk about "breaking the horse" through excessive force to prepare that horse to be ridden or to perform other tasks. This method essentially breaks the animal's spirit so it will submit to its master's will. You might say that what those horse trainers are really doing is breaking the horse's Hara Line. When the horse's Hara Line is broken, the horse has a less "spirited" presentation, almost as if it's sagging or weak, rather than vital and energized. This energetic break brings the horse "into submission" to his master. In some circles, a pimp and his cohorts might repeatedly rape a girl to break her will, so he can more easily control her and sell her in prostitution.

This same kind of breaking of a person using aggressive sodomy also may occur in practices such as hazing in fraternities or sports teams, or in some forms of military training, or to establish loyalty in a gang. It can also occur in repeated childhood sexual abuse as the perpetrator literally destroys the healthy function of the child's body in search of the nourishing life force energy. In fact, the English language reveals the sexual domination mentality woven right into the culture every time someone says "fuck you."

These are all situations where physical and emotional force are used to establish dominance and control. That certainly describes what happens in many rape scenarios. The perpetrator through some combination of violence, intimidation, anger, threats or more subtle manipulation is trying to break down his victim so as to establish dominance and control, not to mention triggering layers of fear or terror. It makes sense, then, that a survivor of sexual assault may feel broken and invaded.

I'll offer one additional image related to the realm of feeling broken by sexual assault. A central concept of Oriental Medicine is Qi, a term I mentioned a moment ago. Qi can be simply described as the life force energy. There are many forms of Qi, and one of them is called the "Wei Qi," which is like a force field or energetic boundary that serves as our body's first line of defense, protecting us from colds and other communicable illnesses and diseases while also shielding us from potentially harmful people and actions that could energetically "get to us." In psychology, we are instructed to maintain healthy boundaries with other people. This refers to the information we reveal, as well as the energy we share with others for the purpose of protecting ourselves and communing with others in safe ways.

Normally, when we trust someone, we take down our energetic boundaries to experience emotional or sexual closeness. But what can happen for many survivors is that their Wei Qi has been invaded, or broken through.

Since they didn't provide consent for the act that will follow, they didn't consciously make their energetic boundaries permeable. When our force field is up and penetration happens, it typically does not feel good. After all, there was a reason the boundaries were up. Something was not energetically aligned; most likely trust and/or love were absent. When trust has been earned, when consent has been given, and we take down our force field, it usually feels good. The screening process or trust building process was successful and we have consciously chosen to merge sexually with the other party.

In sexual assault it just does not happen; the perpetrator simply pushed or forced his way past her boundaries. In a healthy scenario, the "letting in" also would be gradual. Each level of trust gained allows the male into deeper and deeper levels of her soul. Spiritually speaking, we truly are like a many-layered onion. In a healthy scenario, only the highest frequency of love would be permitted into the core of our being, or the inner spiritual sanctum. The high frequency of pure, true, honest love does not pollute or defile. It uplifts and nourishes.

Why is pushing past a boundary so common? There exist degrees of violation with a variety of causes. It could be a desire to cause harm, an indifference to the experience of the person whose boundary is being broken through, impatience or ignoring body language as well as social cues. Also, our fast tempo of life, plus toxic cultural programming that focuses on quantity (of sex partners for a male), rather than the slow trust-building that leads to quality pure love are additional possible causes. The latter scenario is where the true reward lies, but so many are impatient to wait and work for it, thereby defiling the true potential that sexual union holds.

When we bring in terms and descriptions from the field of human energy systems, if you don't find yourself immediately resonating with what may be new language or concepts for you, do not be concerned. A paradigm shift can sometimes seem strange initially but it can also lead to new freedom and understanding.

Here's another way to describe what may happen to us when we're suffering from rape trauma:

Survivors of sexual assault can end up feeling like an "empty shell."

Have you ever looked or felt just totally drained? Have you carried a sense that your whole life force has been used up or taken away? Some environmentally conscious people talk about the practice of over-farming that "rapes the land," meaning that all the nutritive minerals have been depleted from the soil, and the earth's natural ability to regenerate has been disrupted. In some ways, the physical rape of our body can have the same kind of depleting effect.

One term that captures this quality of depletion from trauma is shell-shock. The origins of the term come from war, whereby the trauma of a soldier in the area where an artillery shell exploded and the effect of the blast left him depleted of his vitality, as if it blew out his life force. When you gaze into the eyes of a shell-shocked rape trauma survivor, you see reflected back to you a vacant look. The person is still physically present, but it's like she's not really there. This condition is sometimes described by psychologists as "flattened affect." A survivor in that state may notice that her emotional life, rather than being full of its normal peaks and valleys, is just a flat line—usually a flat low line. The things that used to make her happy no longer move her at all, and she is more and more withdrawn or depressed. She has difficulty experiencing any light in her life, and it often seems as if her very soul has been shattered.

Maybe this resonates in some way with your own experience. Shell shock, also known as Post-Traumatic Stress Disorder—PTSD—is typically linked to post-war experiences and is sometimes extended to a response to major accidents. When we talk about PTSD, we find that the definition and descriptions of what happens to many military veterans also fit naturally with the experience of many rape trauma survivors.

PTSD may include many symptoms: flashbacks; nightmares; repeated memories of the traumatic experience; emotional numbness, with an inability to feel any positive emotions; avoiding anyone and anything that may trigger images of the trauma; being easily startled by stimuli that directly or even indirectly relates to what happened; flashes of anger triggered by people, sounds or situations. Someone suffering from PTSD may also work feverishly to keep people at a distance, while leaning on substances to try to escape feeling emotional pain.

Many rape trauma survivors can immediately relate to one or more of those PTSD symptoms, along with other related signs of major trauma such as guilt, anger, shame, despair or a desire for revenge. We can't say that a rape trauma survivor's experience exactly parallels that of a soldier who served in a war or some other military action. She didn't literally "go to war." For those who took the hit of sexual assault, the war is what happened inside them and the battlefield was their own body. That's where the shock occurred, not in some far-away land. For soldiers, the trauma that rocked their world was delivered by something from the outside and then created internal disorder. For rape trauma survivors, the trauma struck them directly on the inside as a result of the aggressive act committed by the perpetrator against their body and their psyche. And that aggressive act created internal disorder and disharmony.

So we can say that post-traumatic stress caused by rape is in many ways similar to military post traumatic stress. The types of treatments extended to them are also similar. However, as most survivors of sexual assault know well, there are notable differences between the attitudes and perspective toward war-torn PTSD victims and the attitudes toward women who have been raped. Soldiers are more likely to be applauded for their bravery, while rape survivors are often simply not believed because, as we have been exploring, a woman's wounds after sexual assault are more likely to be invisible. It's easy to lose sight of the reality that while war is hell, rape is a war that happens inside you.

So much of what PTSD throws at you as a rape trauma survivor comes in the form of overwhelming emotions and a feeling of having been violated to the core of your being. Even if your sexual assault did not come via outward physical force, but was rather the result of sneaky and manipulative behavior, it's entirely possible that you are experiencing or trying to hold back complex powerful feelings. A different analogy may apply here.

The aftermath of trauma can overwhelm our ability to process powerful negative emotions that may get stuck in our body much like a golf ball stuck in a garden hose.

Many deep emotions can get bottled up inside us after a sexual assault. It often feels as if we've just got way too many powerful emotions to deal with, and they seem out of control or out of balance. A deep sense of grief may dig itself in and leave us doubting whether we can ever move through it. It's the grief that comes with being invaded, humiliated or reduced to that empty shell we just discussed. It is a grief that just emanates from the feeling of having lost something precious to you. In our grief, we may conclude that nothing in our life—our work, our relationships, our daily routines, our sexual being—will ever be the same again. We grieve the woman who was lost in that terrible experience.

Rage can build and lodge itself inside our body. We're just filled with rage at what the perpetrator did to us, and that rage may lash out in misdirected anger. Many survivors will tell you that they have had times when that rage surged up in the form of wanting to kill their rapist. Long ago, in different cultures, that's just what happened. Either the woman killed the perpetrator by poisoning him if he was within the household and she was in charge of cooking, or she had someone else beat him up or kill him. However, that's not the way our criminal justice system works, especially when it comes to rape. Only a token number of assaults are prosecuted and many women feel controlled by fear of violence.

So, what happens after a core level violation when that pent-up rage or terror is not discharged and gets stuck in the body? If not cleared, it ends up eating away at you and after time can lead to medical issues and even disease. Those emotions are kicking and screaming to get out, to have some kind of healthy release. But many survivors fear that if they allow that release, the emotions will be uncontrollable and others may call her crazy, which is a very common way the emotions of women are dismissed in our culture. However, it is the bottled-up tangle of intense emotions that needs to be expressed or cleared or drained for equilibrium to be restored. It is rare that someone in our modern culture talks about "going sane" in this way, by expressing, unfolding and unpackaging the emotional energy within to empty a person's internal world of accumulated pain and horror. In fact, for many understandable reasons, the mainstream medical model is often focused on preventing just this by drugging away, or otherwise suppressing the emotional energy. Or, when it is allowed, it happens in a very controlled and, at times, without an adequate release and resolution within a therapeutic context.

For every action, there is an equal and opposite reaction. Intense energy in means intense energy out—

that is how equilibrium is restored. But when you live in a controlled society based on a very specific social and legal system, whereby you are legally required to prove rape (often very difficult), legally prevented from retaliation (as some animals sensibly do in the wild to destroy predators), you have no choice but to absorb the violence and deal with it as women typically do when other options are not available. Rape survivors in such a society cannot kill the rapist without going to prison. The rapists and abusers often remain free, preying on a multitude of victims. If and when sex offenders are arrested based on testimony only, it takes a large number of victims to come forward to tip the scale of justice in the correct direction, and even then, the victims are often not believed and their motives are questioned. And last but not least, religion may advise victims to forgive, which of course results in a resolution that is incomplete when it is not matched with genuine atonement and restitution on the part of the perpetrator. Who could have invented such a system? Perhaps it was men who accept that they enjoy sexually preying upon others, largely with impunity. Luckily, most men are not like this.

As a survivor of some form of sexual trauma, you probably know all too well the many legal and cultural challenges and injustices we face. However, I trust that by reading this book, you are also deeply committed to your own personal path of healing from this invisible injury. As we will explore more closely in the chapters ahead, there are many safe and effective body-oriented therapies that will enable survivors to clear trauma, release those stuck emotions and bring their bodies, emotions and spirits back into harmony and balance. When I think about how that process can work, I am reminded of what animals in the wild do to release strong emotions triggered by trauma.

Let's say a prey animal is being stalked and chased by a predator. Even if the prey escapes to live another day, the high level of fear from the chase doesn't instantly go away. The emotion can threaten to take its place permanently in that animal's body, leaving its muscles stiff and bound up. But the animal naturally understands exactly what to do to avoid such a fate. It instinctively knows to run off to a safe place, relax for a moment and then begin to tremble. The tremors may even appear to be like epileptic seizures. It trembles and shakes for as long as it takes for the animal to instinctively know that its trauma has been released and it can resume its normal activities—including remaining watchful for the possible return of that predator!

You know that expression when some things happen to us and someone tells us to "just shake it off?" Well, that's exactly what the prey animal does. It shakes the trauma off and frees itself of the possible lasting imprint of the experience. So why can't rape trauma survivors just shake off all those stuck emotions? For a few reasons. First, in the case of the prey animal, the animal was not penetrated to the core as happens in the case of rape. Secondly, being a "civilized member" of a controlled society means that we have been trained from a young age to control our emotions and bodily responses. Nevertheless, there is an excellent body-centered trauma release technique (TRE®) described later in this book that can induce a shaking reflex to resolve some of the symptoms.

For many survivors, the possibility of ever being freed from the intense emotions stuck in their body seems totally beyond reach. Though you came out of the sexual assault, in many ways you may feel as if your soul has been wounded or shattered.

As an acupuncturist, I would describe this depth of emotional trauma as a fragmentation or shattering of your energy flow. The fragmentation means that there are many disconnections and blocks in the overall energy flow in the body. To illustrate this point, I'll provide an example from a minor childhood trauma of mine, which of course is not at all at the level of something as serious as rape; I share it simply to make a point about trauma and how it blocks energy flow.

I was in music class in my elementary school one day when the teacher directed us to line up and take turns singing individually in front of the whole class. That can be scary for a kid, but when I was singing and struck a wrong note, I was met with a heavy layer of shame. "You can't sing. Go to the end of the line," that music teacher said coldly. From then on, I was terrified to sing, even when I was alone. I was traumatized.

Years later, as an adult, I found the nerve to introduce myself to a music teacher who had earned a reputation in our community for her patient and effective teaching of foundational techniques. When I told her about the trauma I was still carrying, she asked me what I wanted to accomplish. I explained that I wasn't looking to go on stage and enter some competition; I just wanted to reclaim the freedom to sing in the shower and not feel embarrassed at how bad I sounded. Six or seven lessons later, my goal was achieved. I stood up in that shower and started singing for all I was worth, and the audience of one—me—thought it sounded pretty darn okay. The block in my energy flow had been released. The golf ball of stuck energy was worked out of the crimped garden hose and the "emotional water" was able to flow freely once again. As I experienced this healing through singing, my experience with acupuncture was similar, in that it helped find the blocks or stagnation in the body's energy system, so the emotional river could flow smoothly again.

So where is that garden hose in our body? In Oriental Medicine, we would say the garden hose is our meridian system, a network of energetic pathways or channels that connect all the organs of our body. The meridian system carries the flow of Qi, which is our life force, and blood throughout the body. When there are specific blocks in that system, disharmony in our body results and various imbalances may emerge.

Different organs are associated with specific emotions in Oriental Medicine. For example, fear or terror is associated with the kidneys. Lungs are linked to grief, which may explain why we may feel a heaviness in our chest when we're very sad. Anger is associated with the liver, and the spleen is associated with worry. The heart is linked to mania, which can result from deep shock or despair. It's a distorted form of the emotion of joy, triggered by an overwhelming event.

If you're looking over that list of particular emotions linked to different organs in your body, you may be saying, "I've got all those intense, overwhelming emotions tormenting me. All my organs must be out of whack, and I'm in serious trouble!" Not to worry. When we go into the details of how acupuncture works, you will learn how treatments are designed to help calm, balance or release all of your intense emotions, no matter where they may be residing in your body.

Emotions can also get trapped in the uterus from rape. Our uterus absorbed a direct hit if we were raped vaginally. A healthy uterus is critical for the emotional health of a woman. The uterus holds many emotions, particularly the frustration and anger that survivors often experience as a result of rape trauma. Often that anger is not expressed; it's stuck. The uterus and the heart are connected by the Chong Meridian or Penetrating Vessel as it is called in Oriental Medicine. In many ways, the uterus is an extension of the heart, and the energetic role of the uterus is to envelop a growing fetus in the mother's love.

So, if your heart feels "broken" by the violation of sexual assault, the energy field of your uterus will likely feel broken. Harmonizing the energy flow in the uterus and healing the connection between the uterus and the heart is a vital part of reestablishing a healthy uterus-heart connection.

Another way to describe the mind, body and spirit wound of a rape trauma survivor is the feeling that something was left behind that we just can't get rid of. Many survivors experience a general state of feeling "unclean" that lingers no matter how thoroughly they wash. That dirty or defiled feeling remains, as if their true essence has been compromised.

Religious traditions often refer to the sexual union of a man and woman in a loving marriage as an act where "the two shall become one flesh," or the husband cleaves unto his wife. The female and male are also referred to in other lore as the chalice and the blade , with the implication that the male sex organ somehow "cuts" into the female. The joining together is more than physical. Each human has an energy flow, which in Qigong practice is called the Microcosmic Orbit. Imagine a line that goes from the top of your head down the midline of your body, to a point between your legs—this is called the Conception vessel (REN). From that point between your legs, the energy flow rises up your back, along the spine to the top of the head, and down the midline of the face—this is the Governor vessel (DU). Energetically, in a sexual act, this energy circuit of the female that is a closed loop is penetrated by the male, and a figure 8 energy flow is created between the man and the woman. In all instances of sexual union between a man and woman, aspects of the spirit are exchanged.

If the sexual union was consensual and based in love, this is a beneficial thing for both parties. He wanted to merge his soul with her and she wanted to merge with his. The loving union uplifts both parties spiritually. But when a woman has not consented to the act and has not willingly and freely allowed the man to cross through her boundary, what gets left behind is unwanted energy that the female either does not want to or cannot assimilate. It will pull her down spiritually. It is an undesirable spiritual part of the perpetrator. Rather than uplifting her being and energizing her spirit, this something can have a crippling effect. It can set up a condition called "Aggressive Energy."

So what is that something that got left behind after you have been sexually invaded? You will probably have your own beliefs about the possibilities, but allow me to share a few ideas.

You could look at it as the perpetrator leaving a piece of his spirit with you, and that spirit more than likely has an undesirable quality to it if it wasn't received with consent or the frequency of love. The perpetrator may have been carrying his own pain from life, perhaps from being the recipient of sexual or physical abuse or other deep wounds that have left him angry, vengeful or hopeless, with a desperate need to gain dominance and control while triggering fear and heartache in the person he attacks. It is also possible that he is someone that typically meets his sexual desires with violence or coercion.

The perpetrator also may be someone who has dabbled in dark occult spiritual practices or who cultivates a dark persona as a member of a gang or criminal group. Or he simply could be someone of a dark or evil nature. That combination of accumulated suffering or negativity that has been absorbed, and the desire to cause suffering, is received by the female as a feeling of having absorbed violence into her inner being and/or a feeling of dirtiness or defilement. Naturally, it can lead to a feeling of low self-esteem as well. And when it's left on you and in you, you may find yourself almost saying something like, "There's something in me that's not me, and it wants to undermine my life or even destroy me." It is an energy that you cannot assimilate—an Aggressive

The Microcosmic Orbit energy flow. The blue line is the Conception Vessel or Ren channel, the red line is the Governor Vessel or Du channel.

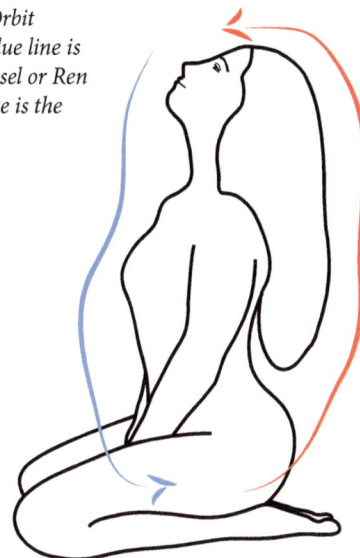

© Morrighan Cox – Lady Raven Art

In all instances of sexual union, aspects of spirit are exchanged,

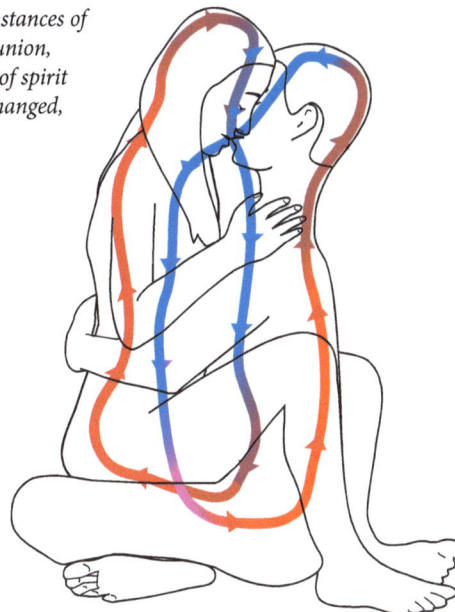

© Morrighan Cox – Lady Raven Art

THE INFINITE FLOW OF GIVING AND RECEIVING
Healthy relationships are about give and take. They are not overbearing, nor are they draining.

Energy condition as it is known in acupuncture theory. It is fairly easy for a man to pull a woman down spiritually by infusing her sexually with hate or pain or violent intent. Historically, this is a common way that a negative manifestation of male dominance was established over a female. The positive way is the male being a true leader, in possession of higher truths and a better plan, that will naturally result in the female willing to follow him. Energetically speaking, our core should be love, and we should receive into our deepest self only love, not anger, not hate, not lust, but love.

Having this uninvited energy in our being is a state that, in extreme instances, might even be called spirit possession or spirit attachment. In such cases, the spiritual energy that came from the perpetrator overwhelms our own spirit to such a degree that we feel that we are not even in charge of ourself. Perhaps you've seen scenes from movies or shows where some demon or entity is depicted as taking over someone's life. The phenomenon is greatly exaggerated for entertainment value, but there are many people in the healing professions today who would say there is an element of truth to the fictionalized experience. Why?

Some spiritual traditions speak of the importance of having everything in your life resolved by the time you die, or "leave your body." The idea is to be clean of any powerful negative emotions, violent tendencies or inclinations, or addictions and other unhealthy behaviors so your soul can go to the next level, or wherever it is supposed to go according to the individual belief system. That's the goal—going "home" to a place of greater love and peace after our life on earth is over. But if a person dies with strong attachments to negative emotions such as anger, jealousy, a desire for revenge, along with addictions to alcohol or drugs, the theory is that an aspect of the soul or spirit is not able to fulfill its goal of moving on. In fact, in Oriental Medicine, the "leave-behind" part of a person's spirit after death is called the "Po."

That piece of a person's spirit may become earthbound, hanging out on the earthly plane because of some strong unresolved desire, or it is simply waiting for people who share the same negative proclivities, who may be vulnerable to its invasion. Maybe you have felt this energy in dreary bars, places where alcoholics or addicts hang out, or seedy brothels that draw people who struggle with lust. If that's the kind of spirit that your perpetrator has picked up and carries, it's going to be transmitted to you in the act of sexual assault. Some esoteric practitioners or healers with a Christian perspective may even refer to it as sexually transmitted demons.

If this all sounds too far out for you, you don't need to wrap your mind around these explanations to understand where that "something" came from and how it wound up damaging your life. If you do live with that sense of being "unclean" after a sexual event, it can be enough to simply conclude that you want it gone. In the chapters ahead, I'll be introducing ways to achieve this important goal in your healing.

Here's another way of defining the invisible injury that causes body/mind/spirit disharmony and suffering:

Depletion

Depletion simply means an emptiness inside. Rape trauma can easily knock out our life force, particularly the energy we carry in our pelvic area or sexual center. What exactly is the "stuff" that has become depleted and how can we get it back? What are some factors that cause depletion in women, and why do young women sometimes experience being "under siege" by men for their sex?

To explain, I would like to introduce you to the term "Jing" as it is called in Oriental Medicine. We've all heard of or worked with the "essential oils" of plants, which possess therapeutic or medicinal properties. Humans, like plants, have essence too. It is the Jing sexual essence.

To help build a broader understanding of this vital essence, we're going to briefly examine a few important social, cultural and gender influences. We'll start with the common scenario of an older person who rejuvenates himself by sexually engaging with individuals who are significantly younger. This practice is one of the primary motivators of sex tourism, sex trafficking, sex with children and also, in many cases, the reason for father/daughter incest. In cases of incest, the wife or mother of the family is typically experiencing depletion, or the husband has sexually

LAST SLEEP OF INNOCENCE
Lily sleeps, her inner flower closed for the night,
protecting her essence.

RED DAWN BREAKS
An act of violence; a trespass against her body and soul
in the most intimate of ways, leaves Lily in shock.

"used up" his wife due to excessive sexual demands, or she simply aged right along with him but he desires the type of sex more typically found in times of youth. So he turns on the daughter, stepdaughter or sometimes even the son for sexual gratification. What the older male is consuming energetically, in this case, is that vital life force essence, Jing. Sex with much younger partners gives the older male vitality of a higher degree than he could get from sexually engaging with his closer-in-age wife. The younger individual is vitalizing the older male by acting as a sexual battery rather than being a partner who also contributes to his spiritual growth. In a balanced scenario, a male and female would actually care about each other and would be exchanging sexual energy with partners who have approximately the same level of vitality, with no extreme differences in age. That healthy scenario results in a loving dynamic balance, which I call the "infinite flow of giving and receiving." But that is not what drives everyone—some are driven by a love of power to the extent where they will draw energy from others for their own benefit.

Virgins especially have a plentiful amount of this Jing essence because it has not yet been tapped into. The 5-to-10 years of sexual relations after the loss of the female's virginity is highly desirable and highly energizing for the male. This is "honeymoon" quality sex, which in normal people does not last into middle and old age simply due to the natural life cycle. This powerful supply of Jing essence is the reason virgins fetch a higher price in the commercial sex industry, and it is one of the reasons that virginity was demanded of brides—the grooms enjoyed anticipating being the first to partake of the virgin bride's magic sexual elixir. Virginity meant that she had a full God-given portion to give her new husband and no other male had partaken of her elixir. That's why the focus on a female's body as being young and vital, versus depleted or "used up," is such a focal point when many men describe

a female. The Jing sexual essence is so desirable, so intoxicating and so vitalizing, that focus on this quality so often eclipses other qualities of the female, such as her character, personality, hopes or dreams.

The old saying "why buy the cow when you can get the milk for free" refers specifically to the female's Yin Jing sexual essence. That is the nutritive "milk" that the female body produces in abundance, primarily during the youthful years of sexual exuberance. A male body has primarily Yang Jing. However, both males and females have a combination of both, with females possessing Jing of a more Yin nature and males possessing Jing of a more Yang nature. Jing is a supreme high, a supreme feel-good and a wonderful rejuvenator. In fact, in Asian cultures it is said that sexual relations with a virgin will even heal a man of serious health problems, it is so rejuvenating. Jing nourishes the brain, the bones, the teeth. However, Jing is not an infinite resource. Just as an orange can produce a certain amount of juice, or a maple tree can give a certain amount of syrup, the Jing essence that a human body can produce is finite.

Life itself will eventually consume our Jing, but that's a natural process that may eventually result in wrinkled skin, sallow complexion, graying hair, decreased sex drive, low energy and weak bones and teeth. Sexual assault, conversely, consumes and depletes Jing via a singular destructive act. It's another stark dimension of the often dark and gnawing reality of the invisible injury of rape trauma. There is so much to heal, but there are many avenues of deep and profound healing available to any woman who chooses to integrate body-oriented approaches.

So, is it possible to restore our Jing essence if it's been depleted? In Oriental Medicine, there is prenatal Jing (the essence you got from your parents), and postnatal Jing, which we get from food and herbal medicines. So, the answer is yes and no. We CAN build postnatal

Jing with tonic herbs, but we cannot rebuild prenatal Jing. If your Jing essence is depleted, build it with safe, Jing building tonic herbs over the long term while abstaining from sex as well as undue stress for a longer period, which could be a year or more depending on the level of depletion. That will help rebuild your vitality. It is possible to learn more about Jing building herbs online (at jingherbs.com as an example); they are safe and have a long history of use. Some herbs are more suitable for women and build Yin Jing, and others are more suitable for men and build Yang Jing. Some Jing building herbs are not specific to Yin and Yang. Some of these amazing herbs are described in the sidebar on this page, and we will further cover this terrain when we explore the many applications of Oriental Medicine in treating rape trauma in Chapter 4.

A survivor of sexual assault may have lost her divine spark.

That spark is simply your divine essence. It's the "piece of God" that you received at birth. In life, we can either cultivate and nurture this spark, or we can squander it through negative behavior, or it can be robbed from us through sexual abuse. One purpose of this book is to help the survivor cultivate that spark back to life.

There are various traditions that teach meditation techniques, prayer, silent worship (Quakers), or Qi Gong techniques (cultivating the three treasures) specifically focused on cultivating the inner light or spark.

Most of us have heard the saying to "let your light shine." No matter how we may name this concept, I think we all appreciate the experience of being around someone who exudes that spark. We've seen people whose eyes are bright and who sparkle with spiritual wholesomeness and goodness, just as we've seen people whose eyes are dim or vacant who have lost their spark. The eyes are the windows to the soul, after all. Again,

Jing building herbs that replenish vital essence

Astragalus
Powerful Qi building herb. Ideal for anyone looking to increase physical energy, increase Lung function, and support immune function.

American Ginseng
This is a health and longevity tonic that anyone can benefit from taking. American Ginseng increases day to day energy, replenishes vitality, and supports Lung function.

Gynostemma, also known as Jiaogulan
Improves health and vitality, builds Qi or life force energy, increases adaptability and immune function.

Eucommia
Increases physical energy and strengthens bones, tendons, and ligaments. Strengthens the kidneys and builds willpower.

Reishi mushroom
Powerful tonic herb for cultivating Spiritual energy, modulating immune function, and promoting health, longevity, and peace of mind.

Cordyceps mushroom
Excellent rejuvenation tonic that is often used by those who are recovering from illness, extended periods of stress, and general feelings of fatigue.

Even if many Chinese herbal formulas are available online, it is best to consult with an herbalist, or talk to the customer service department of an herbal formula supply company such as **Jing Herbs** (jinghebs.com) as they have herbalists on staff. Tonic herbs are generally safe, but it is best to get the correct advice when incorporating herbs into your healing plan.

this is an energetic rather than a physical sense, but it certainly fits the state of many rape trauma survivors. It is clear when the spark is gone. It is a sad fact that there is no shortage of individuals who literally feed on the divine spark of others. Such destructive individuals are sometimes called energy vampires. They don't know how or don't care to cultivate their own inner light so they energetically seek out the light of others to feed on. Cultivating one's own divine inner spark and engaging in regular spiritual housekeeping through prayer, meditation and honest self-review is described by some spiritual traditions as the Right-Hand Path. Energetic vampirism would tend to fall more in descriptions of the Left-Hand Path.

Our female sexual mechanism—the healthy function of female sexuality—is, energetically, like an unfolding flower.

If we're searching for evidence in our culture of women associating themselves with flowers, we need go no further than the many popular female names inspired by flowers: Rose, Lily, Camellia, Daisy, Iris, Heather, Jasmine, Violet. This doesn't mean that women regard themselves as overly fragile or want to be seen that way by others. The link relates more to the way flowers bloom, wilt, go into dormancy, gather reserves, then regenerate before they bloom again. In life there are periods where we are vivacious and radiant, and periods where we go within, to process pain and trauma as we heal and gather enough energy before we go back out into the world and shine again.

Just as we sometimes note how a flower opens and closes, we can say that a woman unfolds her sexual energy in the presence of the right partner, under the right conditions of trust and love. With rape, of course, she has not made the conscious choice to unfold her sexual energy—to take down her emotional and sexual boundaries. The perpetrator invades her force

© Kris Wiltse Illustration

LILY STARTS TO SEE HERSELF AS WHOLE

Lily is finally able to summon the strength to fight the downward spiral into depression and hopelessness. She focuses her mind and imagines the possibility of healing and on the future she wants for herself.

field with an energy that violently tears or crushes the energy field and compromises the healthy function of this sexual unfolding. It's not that different from the idea of "breaking into someone's heart" to get love.

What happens then? The survivor may clamp down on the energy in the pelvic area for fear of attracting any sexual attention at all—this is called armoring. Alternatively, she may feel empty in her pelvic area, a vacuity whereby she has nothing left to give, an emptiness. That helps to explain why survivors often feel as though they are cloaked in a kind of darkness for months or even years after sexual assault. It's a darkness that permeates not only their thoughts but their behavior, their bodies and their spirit.

That's why it's so vitally important, a sacred mission, that if you have been pulled down into the dark tunnels of sexual assault, you learn the best approach and the most effective treatments that can eventually bring you back into the light. When your healing journey moves in the right direction, you can begin to summon the hope that you will live as you are meant to live once more, in your full power and natural radiance.

© Can Stock Photo / tanais

Chapter 2
The Limits of Talk Therapy

For anyone who has been sexually assaulted, the decision to seek therapeutic help often requires a degree of inner strength. It's the kind of courage that many survivors may not believe they can even summon. That's why many women suffering the deep pain that comes from being violated may attempt to file the event away, not talk about it or just wait for "time to heal all wounds." They may decide that it's not really having a significant impact on their lives. Forget it and go on, as they say. Or, if they do choose to confront what has happened, they focus primarily on the challenging and often frustrating landscape of the criminal justice system. The ordeal of navigating the justice system sometimes drains the energy needed to pay full attention to the emotional and psychological damage that the perpetrator left behind.

At some point, however, most survivors will come to recognize that they really do need help, guidance and support to begin to heal. Hopefully, they make this healthy choice soon after the sexual assault. But whenever they reach out for professional help, they will be opening a door that could get them moving toward health and wholeness.

So, whether you are just making that decision now, or you have already tried some form of therapy to address the various symptoms of rape trauma, it's very important to have a clear idea of what may happen during therapeutic treatment. In this chapter, we're going to take a close look at talk therapy and other mainstream therapeutic approaches as they relate to treating the injury of rape. We'll look at what benefits may be gained by working with a traditional or mainstream therapist, and we'll also consider the potential limits of relying solely on this avenue of support for rape trauma. Talk therapy can play a vital role in healing thinking patterns and belief systems that have been negatively impacted by

rape trauma, but in my belief and experience, it can be much more effective if combined with body-centered modalities.

It's helpful to understand that talk therapy, often in the form of cognitive therapy, is the predominant recommended approach for anyone seeking guidance on what direction to follow in healing from rape trauma. This is especially true on the frontlines of rape crisis centers. I learned how true this is through first-hand experience. During my educational training in Oriental Medicine, I conducted a survey in which I asked representatives of rape crisis centers from all over the country and beyond whether they ever recommended body-centered therapies along with talk therapy. Very few of those centers referred survivors to any therapist other than those who practiced some form of talk therapy. The rare exceptions mostly fell into the category of yoga, art therapy or EMDR (Eye Movement Desensitization and Reprocessing). That was encouraging to hear, because those kinds of techniques at least begin to move the trauma out of the body or access the information stored somatically. Most survivors, however, were not even informed of options like these.

Those who direct rape crisis centers either are not aware of the vast spectrum of body-oriented therapies we are exploring here, or for one reason or other they simply don't feel comfortable or familiar with these options as potential therapeutic solutions for the survivors whom they serve. I can understand why that may have been true in the past, but I'm very hopeful that this picture can and will change. I envision the day when rape crisis centers routinely suggest that survivors consider body-oriented therapy in conjunction with working with a trauma-sensitive therapist. Similarly, I also hope that this book may contribute to a sea change in which doctors, pastors, human resource specialists or anyone else who may serve as the first stop for a woman who knows she needs help will also inform survivors of the body-oriented options available to complement talk therapy.

So what kind of help might you find when you turn to talk therapy? All therapists have their own way of working with clients and I certainly can't speak for every kind of experience that a survivor might have in therapy. Generally, though, we can point to many ways in which the process of working with a traditional therapist can potentially assist you in struggling with the wound of rape trauma.

For starters, you are likely to feel less alone. You have someone to talk to, a trained therapist who will listen to you and focus their complete attention on you. That can have a calming or reassuring influence in itself when you are finding that most people in your life don't want to or are not able to really hear you share what you need to say about what happened when you were raped and what's happening in your life in the aftermath.

More than likely, your therapist will assure you that you are not to blame and that the trauma you are experiencing is real. They also may offer a reality check that while others may be telling you that this is something you can just get over, the magnitude of your trauma is such that you can't just will it away. It will take a real commitment to uncover the nature of your wounds and find the best ways to heal them.

Seeing a therapist also can help you sort out and adjust to changes in your daily life triggered by rape trauma. Your therapist might help you identify and develop coping mechanisms for disrupted sleep and eating patterns, flashbacks and nightmares, and social anxiety or withdrawal. Therapists also can assist you in recognizing the new stressors in your personal relationships and help you strategize how to handle them.

The value of simply having a weekly check-in, and a place to be heard and supported, is never to be under-estimated. Many survivors commit to long-term therapy just because they like untangling the many complicated aspects of their life situation.

Many talk therapists, like the counselors at rape crisis centers, will not recommend body-oriented treatment as adjunct therapy. Fortunately, however, there is a growing number of therapists, especially those trained in more recent years, who either integrate some forms of energy work or body work into their own treatment approach or have enough familiarity and respect for body-oriented therapy to mention it to survivors as a potential resource. The landscape for treatment for survivors is slowly shifting, and that means more options and opportunities for you to discover what will help you heal.

So, if you do seek out a therapist who takes some kind of cognitive approach in working with her or his clients, or you are already working with one, you will have many opportunities to make progress. At the same time, it's helpful to be aware of the potential limits of relying solely on talk therapy in pursuing deep healing

First, it's helpful to keep in mind that many survivors who turn to therapy will not contact a therapist and proclaim "I've been raped and I need help." Many survivors focus more on identifying one or two acute issues or symptoms of disharmony in their lives, such as depression, anxiety or poor sleep patterns. They either don't make the immediate connection between these symptoms and the sexual assault they suffered, or they are too uncomfortable revealing that part of their past. So, to get the most from talk therapy, it's important to be open about what happened to you. That will help provide a more solid foundation for therapeutic work. As we've mentioned, there are many potential benefits to be gained in talk therapy, yet there is also the potential to take further critical steps in your healing when you bring body-oriented therapy into your work.

Many healing professionals would say that talk therapy alone simply isn't well positioned to bring out the deeper emotions that may be stuck or blocked in your body as a result of rape trauma. From this perspective, cognitive therapy works primarily with the pre-fontal cortex of the brain where thinking is centered, while the parts of the brain where emotions tend to reside, especially the amygdala, are not typically accessed just by talking.

It should be noted that psychological insights gained during talk therapy can sometimes move or shift trauma held somatically in the body in a positive way. Some therapists may also incorporate deep breathing or relaxation exercises that can begin to contact emotions stored in your body. However, going right to the body in therapeutic work can serve as a more direct and powerful approach in reaching that invisible injury that we talked about.

Remember the golf ball stuck in the garden hose analogy? Talk therapy may resolve matters cognitively, but if the emotions are too intense to express, the golf ball might still be stuck, with the grief, anger or fear blocked in ways that sabotage your life. Crying out your grief in your therapist's office can lighten your load, but it may not take it away.

If you're not directly involving the body in the therapeutic work, you may experience a sense of going around in circles. This can wear down your nervous system, because you're not able to actually release the bound up emotional energy held in your body. Also, in their work with survivors of rape trauma, talk therapists often seek to steer their client to a place of general resolution. In other words, they hope that you will come away from the experience with a basic acceptance that something serious happened, that it wasn't your fault but that you can go on with a "normal" life, that you can accept the changes in your world, that you are better equipped to deal with the day-to-day issues and challenges that you are still likely to encounter as you move forward.

Yet, even if you feel a degree of personal resolution regarding your life as a survivor, you're going to have to navigate life in a culture that is not in resolution—a culture that all too easily condones male aggressive sexual behavior. If you try to go up against all these challenges without a deeper and more complete emotional release and a strengthening of your inner being, you might slip back into despair. You may find yourself saying things like, "It's always been like this for women like me who have been raped, and it's not going to get any better." So keep in mind that even if you do stick with talk therapy and are fortunate to work with a helpful and skilled therapist, you may only reach a certain level of resolution. There is no need to keep carrying such despair. Greater levels of healing are possible for you.

Chapter 3
Why Acupuncture Is a Natural Fit for Treating Rape Trauma

Let's take a moment to reflect on what we have been learning or reconfirming what we may already know about the trauma caused by rape or any act of sexual assault or sexual abuse. Some of the key points we have addressed so far include:

- Rape trauma causes wounds at the body, mind and spirit levels.

- Rape trauma can cause outward symptoms, such as bouts of crying or anger that tilts toward rage, as well as more inward symptoms such as grief and depression, tightness in the chest or gastrointestinal distress.

- Many survivors feel like an empty shell, drained of energy and cut off from their feelings—just barely going through the motions in life.

- The injury of rape trauma is mostly invisible, residing mostly in your body.

- The symptoms of rape trauma in many ways mirror the symptoms of PTSD as they are often linked to those physically or emotionally wounded in battle.

- Survivors of sexual assault often feel as if they have been invaded or broken into, an experience made even more painful with the reality that the perpetrator broke through their barriers or boundaries without consent to gain entry to the most private aspect of their soul.

- The uterus and the heart are closely linked and connected via the Chong Meridian pathway, and a survivor whose uterus has "taken the hit" of sexual assault will likely feel broken-hearted while also experiencing the energetic distress held in the uterus.

- Survivors of sexual assault often carry a sense that the perpetrator has left something behind—some unwanted energy or spiritual component that may be sabotaging their lives and that they just can't seem to get rid of.

- Survivors may also feel depleted and drained, as if they have been looted of their life force essence (their Jing). Moreover, they may feel that the very mechanism of the body which would allow them to regenerate sexual energy, as well as rebuild vitality, has been critically damaged.

If you have survived a sexual assault or were sexually abused while growing up, you probably can identify with one or more of those descriptions. Perhaps your pain and suffering could be defined in other ways as well. Just as sexual assault occurs in many different contexts and situations, with many variable factors, the symptoms of the trauma from sexual assault show up in many ways. The impact may also be felt on another layer of the wound: the frustration or despair that comes from either trying to just get over it, which you discover is not possible, or from diligently searching for the course of healing that will make a real difference in your life but not finding what really works for you.

I understand that frustration and I'm committed to helping you chart a new direction that will get you moving toward a healthier, more harmonious and more balanced life. In these next two chapters, we will explore how Oriental Medicine, which covers not only acupuncture but the broader spectrum of complementary treatment that includes Chinese herbs, may be an especially good fit to fulfill this mission. And we will zero in on some of the ways that acupuncture specifically works in dealing with rape trauma as a treatable injury. Later, I will introduce several other complementary body-oriented therapies that also may be effective in facilitating your healing and restoration.

So why is acupuncture such a useful resource for anyone seeking to gain relief from the many inward and outward symptoms of rape trauma? The short answer is that just as rape trauma affects you on many different levels, acupuncture and Oriental Medicine work on similar levels.

We know that rape trauma impacts you in body, mind and spirit. Acupuncture consistently functions as a directed body/mind/spirit treatment modality. It treats your injury as it relates to the system of organs, which are associated with the many different emotions you may be experiencing. The organs also govern aspects of the soul or spirit, such as the spirit of the heart (Shen), intellect (Yi), willpower (Zhi), courage (Hun) and instinct (Po), and there are acupoints that directly affect these aspects of spirit or soul. Acupuncture works on the musculature and organs that hold on to the trauma memory. Acupuncture frees the flow where energy has become stuck, and supplements energy to where it has become depleted. Broken energy connections in the meridian system are repaired and the body is able to regenerate, balance and heal itself. Another way of defining the multi-dimensional approach of Oriental Medicine is to say that it treats at the organ level; the blood level, which is where the emotions are carried; the Qi level, dealing with our life energy; and the spirit and soul level.

As we go on, you will gain a greater understanding of the many other ways that acupuncture treatments can be seen as working on your multi-layered needs. As we do so, it will be helpful to simply keep in mind that just as rape trauma goes to the core of your being, acupuncture treatment, via the meridian pathways, brings healing to the core.

One sexual assault survivor said that long after her rape, she just couldn't understand why she could not get over it, why it was still in her head. She had not learned that the trauma was not in her head but stored in her body. The reason she could not "get over it" was, most likely, that she still had unprocessed emotions going on inside. She needed a true body/mind/spirit approach such as acupuncture to help lift the dark cloud that still hovered over and around her being. For healing to happen, you simply have to treat on multiple levels. Oriental Medicine does that.

Before I dive into more details regarding the alignment of acupuncture with the needs of survivors of sexual assault, I would like to offer a brief overview of the foundations of Oriental Medicine. This may be especially helpful for those who have not heard much about acupuncture and are curious about how it works.

First, it helps to understand that Oriental Medicine, sometimes called Traditional Chinese Medicine, is thousands of years old. It is a time-tested and sophisticated system of medicine, yet it is constantly transitioning and evolving. Its guiding principle can perhaps best be understood by this ancient Chinese proverb:

"Where this is pain, there is no free flow, where there is free flow, there is no pain."

This philosophical foundation explains the main objective of Oriental Medicine: to release blocked energy in the body, to supplement deficiencies and to drain excesses. The blocked energy often results from holding onto emotions, a common condition triggered by trauma. These energy blocks in your system can affect you in many ways. They can shape the kinds of messages you tell yourself, the choices you make every day and the entire direction your life will follow. Sometimes trauma that was too overwhelming to bear at the time is driven deep into the subconscious and no memory of it can be recalled in the present. However, the individual can simply see that their life is not unfolding as it should. The deficiencies, or depletion, refers to the Qi or Jing—the life force that was lost. The excesses speak to intense emotions that are stored in the body because they are too big to discharge. The goal of treatment, in essence, is to restore that "free flow" to get closer to that state of equilibrium, which would be illustrated by an improvement in physical symptoms and a greater harmony of emotions. When you're free of physical pain and emotional distress, your spiritual harmony can be restored. You become more whole.

Acupuncture has become more widely known and more popular in the U.S. over the last few decades, with millions of Americans now having tried it. Initially, a large percentage of those who chose this modality were primarily seeking relief from chronic pain. That's the area where acupuncture first seemed to break through in our culture, but as patients began to share their success stories of using acupuncture on that front, new doors began to open. More recently, women and men from all circles of life have begun to recognize acupuncture as a powerful healing resource for a much wider spectrum of physical symptoms and emotional issues.

Ted J. Kaptchuk provides an excellent description of the multi-faceted mission of acupuncture in his book The Web That Has No Weaver: Understanding Chinese Medicine. He explains that the very fine needles inserted into strategic points along the acupuncture meridians can "reduce what is excessive, increase what is deficient, warm what is cold, cool what is hot, circulate what is stagnant, move what is congealed, stabilize what is reckless, raise what is falling, and lower what is rising." So as a rape trauma survivor, whether your outward symptoms are evidenced by intense sadness or rage that are running wild, or you are essentially cut off from all your emotions in that empty shell state, acupuncture can treat the imbalance and usher you into a healthier place.

When we have strong emotions trapped in our body, acupuncture treatments can provide a pathway to energetically reduce the intensity of those overwhelming emotions and free that "golf ball stuck in the garden hose." When it comes to feeling like an empty shell, acupuncture can be called on to build back the reserves that have been drained in our body. Specific tonifying techniques, along with the use of carefully chosen Chinese herbs, can bring that energy that was lost back into our body. Our Qi can be freed so that it can supply us with our natural vigor. And if we suffer physical symptoms such as headaches, body stiffness and stomach disturbance, acupuncture can help alleviate that pain because when that blocked energy starts moving freely again, our body will be brought back into harmony.

The meridian system is central to acupuncture treatment. The meridians are bio-energetic pathways that connect the organ systems and circulate life force energy throughout the body. Meridians carry energy, in the form of Qi and blood, all through your body. Emotional trauma can result in energy getting lodged in these pathways, and that energy needs to be released. Or, these bio-electric pathways can develop a type of short-circuit so that the energy is not flowing at all anymore because the connections have been broken due to trauma.

Meridians and points are not visible to the eye, but they can be easily measured by computer aided technology such as the AcuGraph® from Meridia Technology Inc., or any of a variety of battery-operated acupuncture point locator pens, or through classical methods without a machine by an acupuncturist trained in diagnostic skills. A trained acupuncturist knows how to locate points along the meridians that need to be the focus of successful treatment, based on what the person being treated demonstrates and expresses regarding their particular symptoms and overall state of being. An acupuncturist can choose from hundreds of points on the body for treatment. Along the way, the invisible injury of rape trauma is no longer hidden. What can't be accessed visibly begins to reveal itself to the acupuncturist.

Knowing more about some of the ways that acupuncture can effectively address the common symptoms, conditions and problems triggered by rape trauma, we can begin to see why this modality fits so naturally on the path of healing. And rather than regard it as an either/or alternative to talk therapy, acupuncture can be used in conjunction with any therapeutic technique that helps to facilitate more life-affirming thinking patterns and belief systems. Many rape trauma survivors who call upon acupuncture do so while also working with a therapist, which is an excellent way of working on both the body and the mind for a more complete recovery.

Acupuncture is also aligned with treating the most common symptoms of PTSD, such as high anxiety or hyper-vigilance, anger moving toward rage or a desire for revenge, fear of encountering triggers linked to the trauma, increased withdrawal and potential abuse of alcohol or drugs. As we explored in Chapter 1, these PTSD symptoms, more typically associated with military veterans, also appear in the lives of survivors of rape trauma.

In military hospitals, acupuncture has been used successfully in the treatment of PTSD. That is a powerful statement of confidence in the effectiveness of this ancient modality in treatment. If that culture has entrusted acupuncture, used along with other treatment methods and approaches to help soldiers feel less likely to lash out at others and to become less burdened by flashbacks, then any of us as survivors of sexual assault have a solid reason to believe that acupuncture can help address our trauma symptoms.

How can acupuncture survivors who feel broken into or invaded by sexual assault and who suffer the accompanying emotional wound take an important step toward healing with Oriental Medicine treatment? To explain, I will begin by sharing a story from my days as a student training in acupuncture many years ago. Remember when we talked about the Wei Qi, the energetic boundary that is our "force field" that protects us from negative elements or energy coming from the outside? That dynamic came to life for me during one of our training sessions.

We had just started to practice the needling that we would be performing on our patients in acupuncture. At first, we only practiced our needling on oranges. Then, after a few weeks of needling fruit, we began to practice on one another.

Each of us took turns laying down on a treatment table to allow our fellow student acupuncturists to practice on us. We soon discovered that there is a big difference between sticking a needle into an orange and inserting it into a real person. The student acupuncturist would very carefully lean over and first put a stabilizing hand on the practice patient, locate the right acupuncture point per our training, and, as gently as possible, insert the needle. The initial response of these test students was almost always the same: Ouch! In striving to become accomplished acupuncturists who routinely insert needles and witness very little or no discomfort at all in their patients, it was difficult to witness this painful initial reaction. But it taught us something important.

We learned that although we were carrying the best intentions as we approached our fellow student on the table, we didn't have the experience to execute the start of treatment in a way that would be soothing and harmonious for that individual. When we tried to place our hand on our practice patient, it was difficult to

LILY VISITS THE ACUPUNCTURIST
With help from her acupuncturist, Lily's energy pathways are repaired and
balanced. Lily feels peaceful and calm on the treatment table.

convey a sense of confidence and trust because of our lack of experience. As a result, that student kept up their boundary, their Wei Qi. With that armoring still present, the Wei Qi was dense and impenetrable, and anything that penetrated was going to be uncomfortable.

In fact, it is this very dynamic that may be at play when a young man, whose body is coursing with powerful sexual energy, but who also lacks the experience of how to meet his sexual needs with confidence and without causing harm, attempts to sexually penetrate a partner!

Gradually, through trial and error, we became more accomplished at placing our stabilizing hand on the person on the treatment table with a different attitude and manner. Over time, we got much better at conveying that needed sense of trust and confidence. As we did so, the person's Wei Qi, or armor, would become more subtle, and softer, and ultimately just dissolve. Our patient had let her or his Wei Qi down, allowing us to penetrate their skin with our needle without any noticeable pain or discomfort.

A Deeper Understanding

We carry our most private, intimate, precious emotional and spiritual energy within the inner recesses of our heart. The heart, unlike the uterus, does not have a physical pathway to the outside world. To share the intimate emotional and spiritual energy that is deep within our heart, we build intimacy and gradually reveal ourselves to our partner to experience love. The uterus, however, does have a pathway to the outside world via the vagina. Because of this pathway, which was designed to allow for love making and baby delivery, it is much easier to invade. When someone makes contact with our uterus with less than loving energy, the effect is the same as if someone were to enter the inner sanctum of our heart and sully our purest potential for high frequency love.

A secondary part of the problem is that sometimes males are not so good at reading subtle signals and have a mindset that more force is somehow better. Some even mistake the facial expression of trauma for orgasm! The energy of the female body flows in the opposite direction of the male body. When a male enters a female, he is supposed to be making an energy connection and moving his body to start wave-like energetics back and forth like the ebb and flow of the tide. If there is no energy when he enters, she is not turned on. Too often when trauma occurs, the male gets frustrated and decides to use force to try to get to the energy of the female rather than the correct approach, which is to back off and get back to turning her on with foreplay. Much like the process of fracking land for fossil fuel, whereby the earth is violated to the point of ruin to get the energy out, the excessive force breaks the female's regenerative system.

It is important to teach future generations about the heart-uterus connection and how too much force can sometimes result in a "negative feedback loop" whereby the person desiring the energy ends up getting less of it because the system that generates it was injured or destroyed.

When I thought about this experience through the lens of the trauma of sexual assault, I immediately realized that in vaginal rape, the penis is the equivalent of our needle during training. When the penis penetrates a woman with no trust or consent obtained, she definitely has not let her Wei Qi down. The result will be a great deal of discomfort! That pain or discomfort will not go away until it is addressed energetically.

This lesson also enabled me to gain a greater sensitivity and respect for the importance of helping a survivor rebuild a broken energetic boundary wall, the Wei Qi, as part of her healing. Fortunately, there are specific Oriental Medicine techniques to address this need.

Of course, for some survivors of sexual assault, the Wei Qi actually was let down before the sexual act. That could be true, for example, in the case of a pedophile who has been using grooming techniques to build a false sense of trust in the young person who then may initially open to the experience. It also may be the case when a woman has been drugged and is under the influence when the sexual interaction begins. She may have unknowingly let down her Wei Qi, her shield, long enough for the man to penetrate her. However, when the survivor awakens to the reality of what actually happened, and how she very definitely has been violated, the same kind of emotional distress will likely emerge. And she will probably become extremely protective of her Wei Qi in the future. This dynamic is what can lead to the protective mechanism described in the glossary of this book as armoring.

Acupuncture also helps support the healing of the wound to the heart-uterus connection that frequently occurs in the bodies of those who have been sexually assaulted. The uterus, as I noted earlier, can be regarded as an extension of our heart, and it holds many emotions. The uterus is as a very "blood-rich" organ. That's the central area of the menstrual process, and it's where babies are incubated. If a perpetrator's strong negative energy is directed at our uterus through vaginal rape, we get hit with those powerful emotions, such as fear and anger, that the person could be carrying. Several years down the road, a survivor may potentially develop fibroids as a direct result of the assault. So the uterus is an especially vulnerable part of the body for a survivor, and its link to the always vulnerable heart can trigger a serious one-two punch.

In the meridian system, the Chong Mai Meridian actively makes that heart-uterus connection. Treatments focused on restoring and strengthening the energy flow in the Chong Mai Meridian can begin to mend that broken heart experience after being raped.

In fact, treating the Chong Mai after vaginal rape is critical for healing. The intense anger and frustration that resulted from being violated in your most feminine and motherly space can be transformed and released.

Looking more closely at the hit taken by the uterus, and the heart-uterus connection, it becomes clearer how sexual assault delivers an energetic wound. And that wound can be especially deep and traumatizing for girls and young women with limited life experience. Why?

Consider how, just as the uterus and heart are connected for a woman, the heart and the penis are connected in a man also via the Chong Meridian. In an ideal sexual union, both the man and the woman feel a sense of love flowing through those connections. But that's not at all what is happening during sexual assault. More than likely, the male is entering the woman carrying some degree of pain, anger, the need for revenge or a desire to instill fear. Or he may be in a state of loveless lust. That makes for powerful negative energy that will have the effect of pulling a young woman's energy down, especially if the assault is one of her earliest sexual experiences.

We all run energy through our bodies. When you look up at the stage at a rock concert, the rock star you see in front of those thousands of cheering fans is definitely running high energy through his body. You can hear it in the passion coming from his voice and see it in the moves he makes as he dances around the stage. That's an intense energy. In a very different context, a perpetrator is also running intense energy during the act of rape. No woman is prepared to take on that kind of energy, of course, but younger women or children with no or limited sexual experience are much more vulnerable because their sensitivity is higher. Young people just don't have the resilience to carry highly intense bio-electric emotional energy,

The Extraordinary Meridians

In Traditional Chinese medicine, the 8 Extraordinary Meridians are considered to be storage vessels or reservoirs of energy. They are considered energy channels that are formed before incarnation, and guide the process of cell division in the formation of the human embryo.

Row 2 of the illustration shows at what stage of cell division the energy pathways known as the Extraordinary Meridians are formed. The Chong or Penetrating Vessel we have discussed earlier is one of these eight Extraordinary Meridians. These meridians store and distribute our original constitutional energy. As such, traumas that affect us at a core level and thus have an effect on our core constitution can be addressed via treating the 8 Extraordinary Meridians using acupuncture, Acutonics® or other energy medicine modalities.

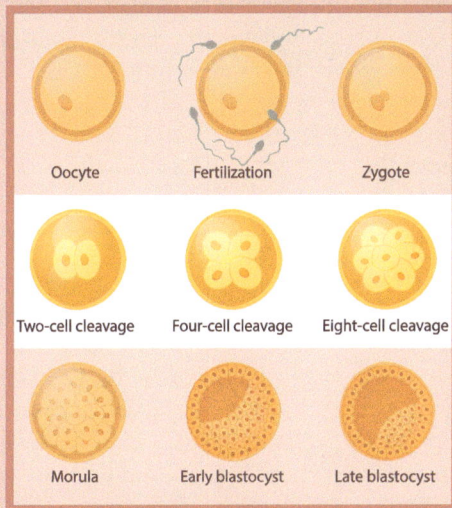

Oocyte	Fertilization	Zygote
Two-cell cleavage	Four-cell cleavage	Eight-cell cleavage
Morula	Early blastocyst	Late blastocyst

© Can Stock Photo / marochkina

The Eight extraordinary meridians are formed at the highlighted stage of cell division

particularly of a negative nature, through their meridian systems. Sexual penetration that causes shock can short-circuit the body's bio-electrical system and create internal chaos.

So what happens when someone interacts sexually with a young person with the intent of gratifying himself without her consent, misusing his strength and power? The hit she takes is all the more traumatic. Her energy may wind up being more strongly affected than a woman who is a proper energetic match to the man. In such a case, you could say that her circuits were blown and her energetic frequency was pulled down and needs to be brought back up. Another manifestation could be the feeling that the soul was shattered. A well-trained acupuncturist would recognize this need, and the treatments that would be utilized would address the young woman's plight with the aim of reconnecting the fragmented flow of energy in the body, and pulling the overall frequency back up.

Many rape trauma survivors turn to alcohol or drugs in an attempt to lift themselves up from the low energetic state of being they were left in after the assault. While it is common to gravitate to alcohol and drugs as a crutch, a crutch is not meant to be used long term. Over the long term, self medicating with alcohol and drugs comes at a tremendous cost to a woman's body and spirit. Survivors who lean on this type of crutch could greatly benefit from knowing that acupuncture patients frequently report feeling high in a natural way after a session. Some patients even describe the acupuncture high as the best high in the world. To feel clear, balanced and light in the body is the most wonderful feeling of all.

Over the course of several treatments, during a period of time when they are also gradually making healthier lifestyle choices, patients become better able to hold that high, clear feeling for longer and longer periods of time.

That's just one of the many ways in which a survivor may be energized through acupuncture treatment.

Some descriptions of Oriental Medicine have compared the system of energy flowing through the body to a network of roads and highways. They explain how anxiety, rage, terror or other intense emotions can act like a traffic jam, jamming up that flow of energy. The resulting traffic backups can spill over onto secondary roads, causing more symptoms to emerge. Acupuncture points, however, are strategically placed along the on-off ramps so treatment can get things flowing fluidly on the highway system again.

Other descriptions compare the meridians to rivers flowing through the body transporting energy—the Qi—to irrigate and nourish the tissues. When the movement through these meridians is disrupted, it's like a dam that can back up the flow in one part of the body and restrict it in others. Acupuncture treatment gets the rivers flowing freely again, thus allowing the emotions to flow smoothly. The names of five acupuncture point classifications reflect this water analogy: Jing Well, Ying Spring, Shu Stream, Jing River, He Sea. As such, acupuncture can be regarded as a system that assists in the proper flow of energy through the bioelectric energy rivers (meridians) in the body.

Another way to further understand how energetic wounding happens is to look at it through the lens of children and their mothers. If a child's mother treats her child harshly through words and actions, it penetrates into that child's being energetically. You don't see the wound, but it surfaces through an emotional disturbance that can shape that child's personality in many negative ways well into adulthood, especially if the wounding happened repeatedly. Receiving harsh treatment from a loved one can feel like getting punched in the gut, betrayal can feel like being stabbed in the back, neglect can feel like an emptiness inside

and so on. Such core-level emotional wounds can remain stored in the body all the while that child is growing up and on into adulthood.

Acupuncture treats this kind of lingering trauma by activating the meridian pathways that go through the impacted area in a way that will finally start moving the energy again, allowing the emotional wound from the mother's harsh treatment to clear. Again, we're talking about a wound or injury that is mostly invisible, like the trauma women suffer from sexual assault. That's part of the multi-level approach that works so well in treating sexual trauma.

Another way in which Oriental Medicine is geared to effectively treat rape trauma is that it integrates the ancient Taoist concept of yin and yang into its philosophy and practice. Yin and yang, which are often referenced accurately or inaccurately in many contexts in our culture today, are two opposite yet complementary energies. There are many ways to define yin and yang. Some simply say that yin is more passive or receptive, while yang is more active. You can also describe them by naming each one's qualities. The qualities of yin are negative, moist, cold, dark, passive and descending. The yin is more closely associated with the feminine. The qualities of yang are hot, dry, bright, active and ascending. The qualities of yang are more closely associated with the masculine.

One of the many ways in which you can see the close link between the concept of yin and yang and the workings of acupuncture is to look at the descriptions of the various organs vital to acupuncture assessment and treatment. The lung, the liver, the kidney, the spleen, the heart and the pericardium, which is the name given to the tissues or membrane surrounding the heart, are all referred to as yin organs. Organs such as the large and small intestine, the stomach, the bladder and the gallbladder are referred to as yang organs.

For survivors of sexual assault, the yin (solid) organs tend to be more prominently impacted. In Oriental Medicine, specific emotions are linked to specific organs: fear with the kidney, grief with the lung, anger with the liver, worry with the spleen and manic inappropriate joy with the heart. Most survivors either struggle with an excess of one or more of those emotions in a way that makes their life unbalanced, or they experience a blockage of those emotions that reins them in and leaves them feeling depleted or empty.

The yang (hollow) organs can also be impacted by rape trauma. The large intestine is linked to a sense of not being able to let go of something psychologically or emotionally, which is something many survivors experience with emotions or beliefs. Acupuncture treatments designed to activate and strengthen the large intestine can help the patient let go of powerful negative emotions. The small intestine is regarded by Oriental Medicine as the organ that separates the pure from the impure in life. So if we had a sexual experience where a degree of manipulation was involved, and we're sorting out whether what happened to us was really pure or impure, acupuncture treatment of the small intestine will assist in that sorting out process. It may sound strange, but these observations are based on thousands of years of practice within the body of knowledge of Oriental Medicine.

The fine needles inserted during an acupuncture treatment are not always placed on the body at locations of these organs. Rather, most often, points are identified by the acupuncturist that connect to those organs via the meridian system, often on the hands and arms and feet and legs.

In regards to the subject of the spirit, entity, etheric slime or "something" that a perpetrator leaves behind in rape, Oriental Medicine also addresses that realm. Older texts on Oriental Medicine actually refer to

© Kris Wiltse Illustration

THE HARA LINE CONNECTS US TO HEAVEN AND EARTH

A healthy Hara Line is not broken, energy flows through the core. It connects us to our higher power, and grounds us to Earth for our life's journey.

these entities as "ghosts," although it's not regarded in the literal sense as we would think of it from stories. I prefer to think of "ghosts" as etheric or spiritual substance, coming from a human. When that spiritual substance is of an extremely negative nature, some religious traditions would call it a demon. No matter what we call that dirty feeling energy that is left behind, Oriental Medicine includes treatments aimed at getting rid of it.

In seeking to gain a better understanding of acupuncture and how it may be effective in advancing your healing process as a rape trauma survivor, it can also be helpful to recognize that Oriental Medicine differs from Western medicine or mainstream psychological treatment in that it does not operate from a basic cause/effect model. Oriental Medicine does not seek to isolate one symptom or problem and then track the specific cause behind it. Rather, the goal is to gain a much bigger picture that encompasses not only our specific symptoms of trauma but our whole life situation. And rather than calling upon just one treatment that's going to "fix" the problem, acupuncturists are likely to proceed with several treatments aimed at different dimensions of our imbalance and disharmony. Again, acupuncture works simultaneously on several levels.

While you are being treated, you also are having a much different physical and mental experience than you would in mainstream or talk therapy. As we touched upon in the previous chapter, with a body-centered therapy like acupuncture you don't have to fully re-experience your trauma on an emotional level by talking it out or crying it out. You are comfortably lying still on the table, with no words needed, while you are receiving the treatment. While on the table, it is best to get into a meditative state that is halfway between sleep and wakefulness where you actually feel your spirit body hovering slightly above your real body just as it typically does right before you normally fall asleep. It is in this state that the healing happens!! The acupuncturist will likely have soft relaxing music playing and will dim the light and leave you alone while the needles free the body's energy flows and the healing takes place. There may still be a time and place for you to get back to talking about your trauma, but when you do, the intensity of the emotional charge is likely to be radically diminished after you have been treated with acupuncture.

With that "free flow" that acupuncture strives to achieve, you will be much better equipped to deal with any situation in your life. The grip of trauma will be loosened, and that divine spark will gradually return.

Chapter 4
Resurrecting the Spirit

No matter which of the many effects of sexual assault you may be suffering as a survivor, it's very likely that somehow or other your life has negatively been altered since that traumatic experience. Spiritually, you have been catapulted to a lower level. Something vital has been lost. What many would call that divine spark may be gone.

Only you know the specific details that add up to the absence of that spark for you. But if you make the decision to pursue Oriental Medicine to try to get back what may be missing, and to heal what feels broken, I can assure you that you will be opening a door that could potentially lead you toward reclaiming what you lost.

In the previous chapter, we touched upon some of the basic reasons why Oriental Medicine can be helpful and effective in facilitating your healing. Now we're going to focus a bit more closely on how Oriental Medicine specifically works in the treatment of rape trauma. I'll describe what you may expect in an acupuncturist's treatment room, and I'll explain the different approaches that your acupuncturist may take in finding what will work best for you in your journey towards healing.

Please do not feel concerned about any need to become an expert on the inner workings of acupuncture and its treatments related to sexual assault. You don't have to gain any particular body of knowledge of Oriental Medicine in order to reap its many benefits in addressing your physical, emotional and spiritual needs. Your acupuncturist is the one with the training. All you have to do is to show up, engage fully in the intake and assessment process, relax during your treatments, and do your best to execute the follow-up plan that your acupuncturist will suggest.

I'd like to offer a special word of encouragement to any survivor who may feel anxious about trying this ancient, but to some people, strange therapy. My hope is that reading this chapter will help you feel more comfortable about beginning therapeutic work with an acupuncturist. If you're still looking for a way to ease into these new waters in an approach that is especially cautious, you may want to consider asking a friend to volunteer for an acupuncture session while you just watch. Or, if you are uneasy being in a small room with anyone, trying something new, it is always possible to have a friend or relative come with you, and have them sit in the room while you receive your first treatment.

All medicine, including Oriental Medicine, is an art as much as a science. If there is more than one acupuncturist in your community, I urge you to take your time in interviewing potential helpers to select the practitioner that feels right to you. You should definitely feel comfortable with that person before you make a commitment.

It may be helpful to know that before your acupuncturist can begin to work with you, they need to complete a detailed intake or assessment process. Through some combination of written or verbal questions, they will be asking you to provide information related to many areas of your life. This is likely to be a different experience than meeting a Western physician or psychologist where they may begin by simply saying, "Tell me what the problem is" or "Why are you here?" Those practitioners may be operating from a belief system in which healing is simply about zeroing in on the one "problem" and selecting the one primary treatment option to "fix" it. In the acupuncture assessment process, the questions that you will respond to will cover the physical and emotional distress you identify as your primary problems or issues, but they don't end there.

Before I go on to outline some of those other questions, this is a good time to make an important point about the experience of rape trauma survivors. It is true that many survivors of sexual assault either do not recognize the direct link between their current suffering and the sexual assault they experienced, or they may be too hesitant to reveal that part of their life right away. In no way does this mean that they can't receive acupuncture treatment that will effectively address what has happened in their lives because of rape. Oriental Medicine tracks the disturbances and disharmony in your body to determine where your energy is in disharmony, and treatment is oriented to free those blocks, calm the disturbance and restore your body and emotional state to harmony. It is not critical that this happens under the clearly identified umbrella of "rape trauma" or any particular symptom you are able to identify. Of course, even when a survivor does not immediately mention having been sexually assaulted, that information may organically be revealed during the intake process.

The questions you will be asked will most likely survey your broad health history, noting any significant events all the way from childhood. The answers you provide will help your acupuncturist develop a deeper understanding of your overall mind-body state and how to begin to restore harmony. The reality is that few if any survivors arrived at that experience as a perfectly whole and balanced person anyway. Maybe you've wrestled with a chronic health issue, or there were issues with a pregnancy. You may have been abusing alcohol or drugs, or perhaps you suffered bouts of insomnia even before your trauma from sexual assault further disturbed your sleep routine. The potential that emerges through Oriental Medicine treatment may have a positive effect on issues that go beyond what came about as the result of rape trauma.

Questions about a person's sleeping patterns are commonly asked during the intake process. At some point when your acupuncturist is assessing what you reveal about your sleep patterns, he or she may be utilizing a diagnostic tool referred to in Oriental Medicine as the Circadian Clock. To illustrate how this works, let's say that you revealed that you frequently wake up angry between 1 a.m. and 3 a.m. The Circadian Clock identifies the nature of the disturbance that shows up during that period. In this case, the liver, which governs anger, is overwhelmed. If you report being disturbed from sleep between 3 a.m. and 5 a.m., your acupuncturist will be able to determine that your lungs need support and you may be holding on to excess grief.

Other questions asked during the assessment will steer you toward digestive issues you may be dealing with, along with the general frequency of elimination. As a woman, you also will be asked to share information related to past or present pregnancies, as well as your menstrual period. Additional questions about your lifestyle also may come up.

It's helpful to answer all these questions as openly and honestly as you can. The picture that emerges through this process will greatly assist your acupuncturist in deciding what course of treatment to follow. When you talk with your acupuncturist during the intake process, that person also will be noting useful information from your speech patterns, emotional state, subtle color to the skin, even subtle odor emanating from the body. All of these are diagnostically relevant. Again, that's all just useful background for this practitioner to guide the diagnosis. As practitioners of Oriental Medicine, we are trained to cultivate a diagnostic eye, ear and touch.

During this process, your acupuncturist also will take time to read your pulse by gently placing three of his or her fingers on each of your wrists. This process is different from how a Western medicine doctor or nurse takes your pulse to obtain your heart rate during a basic examination. Your acupuncturist is actually listening to your pulse because in Oriental Medicine there is a six-point pulse system that can help to pinpoint certain disorders and decide upon the most effective treatments for them.

Your acupuncturist also may ask to observe your tongue. When you open your mouth in this process it's not the same as sticking out your tongue and saying "ahhh" in a Western medicine doctor's office. Instead, an acupuncturist is observing the color, shape and coating on your tongue. Again, your acupuncturist is simply looking for signs that reveal information about the condition of your various internal organ systems. You would be surprised at the depth of information that the tongue reveals.

One way you might view this assessment process is to remember our exploration into how rape trauma is an invisible injury. The practitioner conducting this intake and observation has begun to do the important work of identifying what may seem "invisible." As an example, when your acupuncturist was listening to and reading your pulse, she or he may have been identifying a specific organ system that is out of balance. The pulse on each side reveals particular qualities that are linked to those organs in your body. So, if you are frustrated, or full of rage, the pulse in the liver position would be very wiry and strong on the mid position of the left wrist. If your lung pulse is weak, in the distal position on the right wrist, the acupuncturist would be reading a state of grief or depression and charting a course of tonifying your lungs during treatment.

Sometimes during the course of reading the pulse, an acupuncturist identifies a condition called the "Qi Wild" pulse. Basically, that means the person's pulses feel like chaos, and their psychological and emotional state is disordered or chaotic. That happens to be true for many rape trauma survivors, but it is not in any way a sign that we have insurmountable problems. It's simply an indicator, or a confirmation, that we are dealing with many intense emotions that are undermining our well-being. The treatments our acupuncturist will zero in on will be geared toward addressing our total state.

After the detailed intake, you will lie down on the treatment table and very fine needles, usually between five and twenty of them, will be inserted in various points on your body, which may include your hands, your back or your ear. The needles are inserted at different depths depending upon the location of the points and the level the acupuncturist is treating. Remember, the location of the inserted needles is often not in any place near the actual impacted areas of your body. The places selected for needle placement are linked to the meridian system that channels energy to those various organs and impacted areas of your body. For example, if your rape trauma has left you with a feeling of congestion, or dense dark energy in your pelvic area, it is the Liver Meridian that flows to the groin. Treating points on the foot on the Liver Meridian activate a river of energy that frees the flow of that congested energy in the groin.

In Oriental Medicine, the meridian pathways are the carriers of both Qi and blood. Ted J. Kaptchuk, acupuncturist and herbalist, describes them this way in The Web That Has No Weaver: "Qi is the quality of life that is dynamic and transforming. Blood is the responsive, accepting, effortless, soft, and nurturing complement of the clinical Qi. Qi and Blood are the Yin and Yang of ordinary life activity. Qi activates, Blood relaxes. Qi quickens, Blood softens. Qi is tense and tight; Blood is smooth and languid. Qi embodies effort. Blood is effortlessness. Qi is becoming. Blood is being…. Qi creates and moves the Blood and also holds it in its place. Blood in turn nourishes the Organs that produce and regulate the Qi." Both Qi and Blood need to be harmonized in the body for emotional health.

In Chapter 1, we introduced some of the key connections between specific organs and emotional states that you may be experiencing as a survivor: kidneys linked to fear; grief and sadness connected to the lungs; anger associated with the liver; worry linked to the spleen; manic, out-of-control joy connected to the heart. The acupuncture points chosen for a particular treatment are often directed toward one or more of those impacted organs. Again, when you are being treated, you would not be aware of which organs are being targeted by the placement of the needles. You really don't need to know, although acupuncturists are generally more than happy to explain if you are curious. Your work as a patient is simply to relax and allow the treatment to happen.

Emotions, after all, are energy in motion. It's possible that you may experience an emotional release during a treatment. The needles move the Qi and blood in the body to mobilize, transform and dissipate the emotions. Therefore, an emotional release is a natural and healthy occurrence if it happens. On the flip side, if you do not experience any particular emotions and simply feel so calm you want to drift off to sleep, that is in no way a sign that important emotional healing is not happening. The work of acupuncture on your body's energetic system is subtle but it is also profound. Some patients report that the movement of energy during a treatment can unfreeze frozen memories held in the body. In such cases, a therapist could be on hand (perhaps scheduled after the acupuncture treatment) to talk through what has surfaced.

When we explored the meridians earlier, we mentioned the importance for rape trauma survivors of the Chong Mai Meridian or Penetrating vessel, which connects the uterus to the heart. This is a vital meridian to treat because of that broken-hearted state survivors often experience and because the uterus, such a vital organ for any woman, holds many emotions disturbed by rape trauma. There is another piece of the healing equation involving the Chong Mai Meridian. It shares

points with the Kidney Meridian, and the kidney is significant because the kidneys govern sexuality as well as the emotion of fear.

During a sexual assault, many survivors feel an internal snapping, as if some connection was broken. What snapped or broke may have been the Heart/Kidney axis. This is a major blow to a woman's body and being because after it happens, she will become far less able to feel any warmth in her heart during sexual union. So, if you have survived a sexual assault and months or years later, you find yourself telling a male partner in your life that you just don't feel the intimacy of sex you once did, it's not just a matter of not "getting over" the rape. There was a very specific injury that you suffered when this vital energetic connection was broken. Understand it as a short circuit of one of the body's bioelectric pathways or meridians. After that connection is restored and healed as part of your comprehensive treatment in Oriental Medicine, you will gradually find that your ability to experience physical warmth and closeness during sex with a partner with whom you share mutual respect and love will return.

You may also recall from our earlier discussion how our basic, first-line-of-defense energetic boundary called the Wei Qi may have been forcibly penetrated during sexual assault. The strengthening of the Wei Qi during acupuncture treatment, or with an herbal remedy, is often a critical component of the healing of a survivor. Once our boundary is restored, we no longer feel inclined to let any potential sexual partner into our body because we don't believe we can fend off any possible penetration anyway. Nor do we have to isolate ourselves and avoid any and all potential sexual contact as a way to ward off a new invasion. With a restored Wei Qi, we experience a greater balance and harmony, which allows us to make decisions regarding sex and intimacy from a healthier place.

Another energetic boundary often takes on added significance for a survivor. This boundary is maintained by the Pericardium, which in Oriental Medicine is the protector of the heart. A woman who has been sexually assaulted often experiences stagnant Qi in the Pericardium. As a result, energy is not able to reach the Heart. In Oriental Medicine, this state is referred to as "Heart Closed," and it can lead either to a heart that has grown cold, meaning the person has lost passion for life, or to a heart that has too much "heat," resulting in palpitations, insomnia, talking too much, etc. If your acupuncturist diagnoses stagnant Qi in your Pericardium, treatment will be directed toward restoring the correct function of the Pericardium. When that treatment is successful, your heart will once again be able to receive nourishment and thus not grow cold, and that heat or pent-up energy can be released so you no longer will feel inclined to lash out at others in present situations because of the injury you suffered in the past. That's just one more way that a highly trained acupuncturist who tunes in to conditions that may be related to rape trauma can help facilitate your healing.

The Clinical Practice of Chinese Medicine by Lonny Jarrett is an excellent resource. The author is a scholar and a well-established, highly respected practitioner of Oriental Medicine. Though his writing approach may be a bit too technical or academic for most rape trauma survivors, acupuncturists and other healing practitioners may benefit greatly from his accumulated knowledge, wisdom, insights and hands-on information.

Acupuncturists also may utilize what is known in Oriental Medicine as the Eight Principles of Diagnosis at some point in deciding how to focus treatment. The principles are actually four sets of opposites: yin/yang, internal/external, cold/hot, deficiency/excess. For example, your acupuncturist will assess whether your issue or pattern of disharmony is a condition going on internally, or inside your body, as would be the case for many core emotions, or whether it may be something going on externally, such as a physical blow suffered during a violent sexual assault. Are you dealing with cold, like the cold and closed heart we just mentioned, or is it hot, as evidenced by that urge to lash out in anger or an unstoppable outpouring of tears? Is your condition more yin, the receptive, or is it more yang, the active? The evaluation of deficiency or excess may certainly relate to the spectrum of your intense emotions: are they blocked like that golf ball in the garden hose, or are they running out of control in a way that makes it almost impossible for you to feel at peace?

Interestingly, many survivors of sexual trauma are both deficient and in excess. How? Well, you may be having that sense of your life force being knocked out of you. That's deficiency. At the same time, you could be experiencing a great deal of anger or anxiety about how this event has derailed your life. That's the excess part of the equation. Your acupuncturist will be able to find ways to address both conditions, because this combination of excess/deficient can surface in someone who did not suffer major trauma. In Oriental Medicine, there are specific treatments for "wired and tired." Patients may be wired due to an emotional and mental state of hypervigilance, while in actuality they are hiding an internal deficiency. Interestingly, the fast pace and various stresses of modern life in and of itself can lead to the condition of being wired and tired.

With any of these approaches to treatment, your acupuncturist will be devising what we call a point protocol. This term simply refers to a selection of needles that are placed on specific points with a very specific intention or desired outcome, consistent with the many different states of distress and disharmony. Since Oriental Medicine is so ancient, many of these point protocols have been proven effective over a considerably long period of time. To give you one example of just how precisely some point protocols are framed, there is a point protocol specifically for feeling rage to the point where you want to kill someone. It's certainly not unusual for some rape trauma survivors to experience that level of rage at some point.

Another specific point protocol that has proven to be deeply calming for PTSD and therefore can be effective in treating rape trauma is the NADA protocol. As I mentioned, the fine needles used in acupuncture may be inserted at many different places on your body as connected to your meridian system. In the NADA protocol, all the needles utilized are placed on your ear. In Oriental Medicine, the ear is considered a microcosm of the entire body, in the same way that the foot is considered the microcosm of the body in reflexology. One point of this five-point protocol is the Shen Men, which is sometimes referred to as "The Gate of Heaven." The purpose of Shen Men is to tranquilize the mind and to facilitate a state of harmony and serenity. This master point alleviates stress, pain, tension, anxiety, depression, insomnia, restlessness and excessive sensitivity. Putting pressure on this point is believed to be a way to bring celestial energy to your whole body. Shen is the term for spirit of the heart in Oriental Medicine.

If you were to superimpose the fetus in the womb on the ear, the precise location would be the place where the points in this point protocol are centered. So the NADA protocol is one more potentially useful treatment approach for anyone who has survived a sexual assault.

Another auricular (ear) point that holds great promise in the treatment of survivors is called Point Zero. This master point is the geometrical and physiological center of the entire ear. It brings the whole body towards homeostasis, producing a balance of energy, a balance of hormones and a balance of brain activity. It also supports the actions of other points.

If your acupuncturist informs you during one treatment session that some needles will be placed on your ear that day, you now have a better understanding of why that choice has been made.

As survivors, we may be well aware that the deep wounds we suffer from this mostly invisible injury often take the form of intense emotions that some survivors may have managed with drugs and alcohol. We may have deeply cried in our pain and loss, we may have screamed at other people for reasons we can't even explain, we may have awoken in the middle of the night with rage or high anxiety, or we may have felt the terror of facing people and situations that remind us of what the perpetrator has done. Many other emotions may also be stirred in a level of intensity and frequency that we can't imagine bearing much longer.

You can be confident that your acupuncturist recognizes and has identified those intense emotions and the cumulative harmful effect they have had on your body, as well as your mind and spirit. Of course, if you are used to being treated by practitioners of Western medicine, where one central ailment or disease was diagnosed and one primary treatment approach or medication was called upon to take care of it, you may be hoping for that "one magic cure" for everything that's happened to you as a result of sexual assault trauma. Unfortunately, that's not possible in the treatment of this kind of far-reaching, multi-level wound. However, there are some approaches in Oriental Medicine that seem especially well suited for addressing a great deal of your pain and suffering. As it relates to those intense emotions and the havoc they cause, one of those acupuncture treatments is called "Draining Aggressive Energy."

The name of this procedure does not mean that your acupuncturist has assessed you as an overly aggressive person. In Oriental Medicine, we're talking in terms related to emotional energy of a harmful nature that is held in the organs. Draining Aggressive Energy simply refers to addressing the "aggressive" or intense emotions energetically linked to specific organs in your body. Aggressive energy in the liver is rage, Aggressive Energy in the kidneys is fear or terror, Aggressive Energy for the lungs is deep grief—like that sinking feeling in your chest when someone you love dies or a part of you has died—Aggressive Energy in the heart is that manic or distorted joy originating from deep trauma or shock that we mentioned earlier, and Aggressive Energy in the spleen is constant worry. To clarify, the spleen as it is known in Oriental Medicine is not the same as the spleen as defined by Western medicine, but rather the pancreas and its function in Oriental Medicine that goes far beyond Western understanding.

Your acupuncturist could utilize the Draining Aggressive Energy treatment with the recognition that you may well be experiencing aggressive levels of emotion related to more than one organ. So rather than relying on a process that only pinpoints one organ and its associated emotion, the practitioner has an opportunity to focus on all of them at once. This is especially effective because these are the emotions often associated with any trauma, including the trauma resulting from sexual assault.

For this process, the acupuncturist will be inserting those very fine needles on points on your back. The needles are inserted superficially, barely breaking the skin, and they are placed on the points on the bladder line that directly connect to the impacted organs. If Aggressive Energy is found to be present in those organs, something interesting happens: it will rise to the surface of the body and a red circle will emerge around the specific needle. The presence of the red is a confirmation that the Aggressive Energy, representing emotions trapped in the organs, has been given a pathway out of your body. Eventually, the red circles

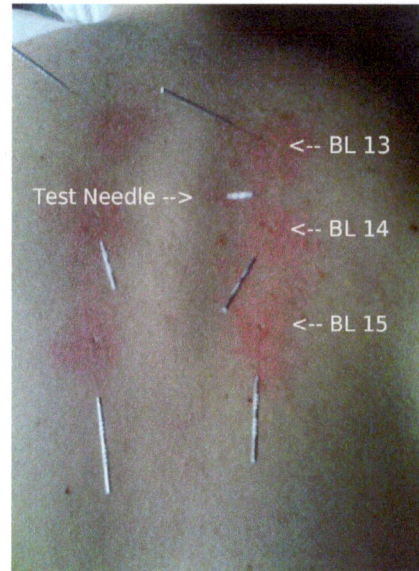

*Photo courtesy of Michelle Gellis M.Ac. L.Ac. Acupuncture

diminish, leaving behind a sign that the Aggressive Energy has been drained.

Needles are left in the patient until the redness that appears around the points is gone. A test point can be added to this treatment to gauge the redness appearing at specific points in the treatment.

The desired result, of course, is a healthy change in your life. After the energy is drained, you are likely to experience less intense fear, grief, anger, worry and mania. This can be a welcome shift for anyone who has felt engulfed by emotions that wear you down and leave you frustrated, or even hopeless about your present and your future. You just feel calmer, lighter and more clear. And for any survivor who had been swallowed up by trauma, the change provides the opportunity to react to present-day situations without being triggered by emotions that are specifically linked to being sexually assaulted. That may mean that you become less reactive

Draining Aggressive Energy Treatment

Short needles are inserted just barely piercing the skin, from top to bottom and right to left, usually inserting BL-15, the heart points, as the last point.

The points in this treatment are:

BL-13 Lung Shu
(associated emotion: grief)

BL-14 Pericardium Shu
(heart protector)

BL-15 Heart Shu
(associated emotion: mania)

BL-18 Liver Shu
(associated emotion: rage or anger)

BL-20 Spleen Shu
(associated emotion: worry)

BL-23 Kidney Shu
(associated emotion: fear or terror)

Since sexual trauma usually results in a combination of these strong negative emotions, the Draining Aggressive Energy Treatment is an ideal choice, particularly soon after the event, or even in later stages when strong negative emotions surface.

It's a very effective protocol for harmonizing the emotions and **I would consider it an EXTREMELY important treatment for all sexual trauma survivors**, to be used at the beginning of a series of treatments or several times throughout a course of treatment if more body memories come up.

when triggered and thus you gain a major level of freedom to be who you are and make choices that you actually want to make, rather than being forced into choices because you're caught up in the grip of one or more intense emotions.

This process, as executed in acupuncture, may appear to have some similarity with what happens if you are being treated by a body energy practitioner or massage therapist, or by practicing yoga. Imbalanced energy is being addressed in your body. The difference is that acupuncture is able to drain toxic or pathological energy from the body much more directly. With the Draining Aggressive Energy treatment in acupuncture, the needles actually allow for a pathway directly over the affected organs for harmful energy to exit the body so that your emotions settle into more manageable levels.

As an acupuncturist, I have used the Draining Aggressive Energy treatment on patients, but I have received it as well. One day many years ago, someone I loved and respected misinterpreted a situation, flew off the handle and hit me in the face. You could say that my buttons were pushed, big-time, and I felt anger and shock as this event was completely outside the norm of my everyday experience. That day, I happened to be in the company of a fellow acupuncturist whom I knew and trusted. "I really need the Draining Aggressive Energy treatment," I told her, knowing that she would be able to help me. Midway during the treatment, I requested that the acupuncturist take a photo of my back with the needles in. When she showed the photo to me, sure enough, there were those red circles prominently displayed around the needles. The Aggressive Energy was exiting my body. The Aggressive Energy in this case was the shock, anger and insult of having been physically assaulted by someone I trusted. Thanks to this treatment, I was able to resolve the situation from a higher place, as the better person, cool and collected rather than angry, retaliatory and out of control.

Because a survivor often absorbs aggressive, violent or negative energy into the core of her being during sexual assault, the Draining Aggressive Energy is an excellent protocol to use in the treatment plan.

Draining Aggressive Energy also may be called upon for those who have had some kind of interaction with a person whom they may experience as "unclean" in an emotional or spiritual rather than physical sense. The recipient of this contact with the "unclean" person may feel an internal conflict because that other person and her or his behavior are far out of alignment with the recipient's spiritual nature. You could say that the recipient got hit with some "bad Qi" and the Aggressive Energy, as evidenced by suddenly intense emotions, would need to be drained before he or she would be able to feel in balance and harmony again.

That description certainly fits the experience of many rape trauma survivors. They were hit with bad Qi carried by the perpetrator, and this resulting Aggressive Energy has probably been exerting negative influences on their lives ever since. If that's happened to you, that negative influence simply has to be cleared before you can be yourself again. Draining Aggressive Energy may be one important approach in making it happen.

As part of understanding how Oriental Medicine can be effective in facilitating your healing, it's helpful to learn more about the general process of working with your acupuncturist over the course of several treatments. Those treatment sessions, which include relaxing with the inserted needles on your body for approximately thirty to forty minutes, are not the only part of your experience working with Oriental Medicine. As a complement to their hands-on treatment, acupuncturists often prescribe Chinese herbs, which we will discuss in a moment. If you are also working with a traditional therapist at the time that you begin acupuncture treatment, your acupuncturist may offer guidance on how to continue to best utilize that resource. Your acupuncturist also may recommend lifestyle changes related to diet, exercise and other routines designed to support the changes that Oriental Medicine are helping to promote.

In my own practice, I have found that rape trauma survivors or even other types of patients who are going through any type of deep healing from a soul wound can benefit from playing soothing sacred music at home at night during sleep at low volume, set to repeat. The body restores and repairs itself at night and the vibration of sacred sound infuses the body and spirit with healing. I stress choosing ancient and sacred music that truly moves a person's soul. To give a few examples, Tibetan chanting is especially effective at moving my soul. Vedic chanting, which is centuries-old Hindu music, and the chanting of Benedictine monks may have a similar effect. Music that soothes our emotions contributes to the healing going on during acupuncture treatment. Most of us instinctively call upon some kind of music to help us feel better at different times in our lives, but for this purpose you will want to choose music that's gentle, with natural instruments and natural voices. Why sacred? When feeling defiled, bring in the sacred at every possible level, including auditory and yes, even, and especially at night. Infuse yourself in sacredness to counteract the darkness.

The next acupuncture treatment that I'll describe picks up on a subject we touched upon in Chapter 1. Many survivors of sexual assault come away from the experience with the sensation that something unwanted and harmful has been left in their body by the perpetrator and they don't know how to get rid of it. This "something," which can be called by many names, is almost like having some kind of etheric muck, spirit fragment or etheric parasite that has invaded your system. In extreme cases a person might even describe it as a demon.

This unwanted etheric remnant, as we mentioned earlier, is sometimes referred to as spirit attachment or spirit possession. There have been many approaches referenced as being effective in getting this kind of "bad spirit" out of a person's body and being. William Baldwin, in his book Spirit Releasement Therapy, presents a fascinating process for removing that unwanted entity by bringing the patient under through hypnotherapy and guiding the negative entity into the spirit world. He details specific steps, which include identifying the entity, dialoguing with it to discover its purpose and needs, and releasing it into the light. The desired result of this releasing process is to free the person of unwanted influences, which for a survivor of sexual assault may have taken the form of self-destructive behavior or the manifestation of character traits that are not of the self and most times hijack the host's true destiny.

That's one approach to liberating someone from that unwanted, parasitic type of entity or influence. Oriental Medicine has its own way of addressing the problem, and various schools of acupuncture have identified this area as critical in the healing process for a survivor of sexual assault.

Acupuncturists who set out to treat this issue bring a broad understanding to the equation. In Oriental Medicine, we talk about the different orifices that serve as gateways to the inside of the body: the mouth, the eyes, the nostrils, the ears. We've all heard the popular expression of the three wise monkeys: "see no evil, hear no evil, speak no evil." That's a basic urging to keep ourselves, in particular our core, as pure as we can, to avoid those people and actions that can corrupt our souls and undermine our attempts to lead a healthy or spiritual life. By extension, we can also infer that fully and openly engaging with those who have not adhered to these principles and who are living an impure or dark life can corrupt our purity. Of course, in today's world

these "evil" influences are lurking seemingly everywhere and can be easily picked up. In fact, I am convinced that the proliferation of Internet pornography has served to not only corrupt the souls of those who partake in it but to weaponize individuals with low impulse control against the innocent. If we don't become aware of what's happening, dark influences can begin to cultivate a more evil or darker nature in us. But as the well-known saying reminds us, we have choices about we do to either embrace or reject such darker influences.

Rape trauma survivors, however, are denied one aspect of that choice by the act of sexual assault. It's important to recognize that those orifices that serve as gateways to our body and can profoundly impact our being also include the vagina and the anus. So if a perpetrator whose soul is of a dark or even evil nature invades any of our orifices, he is partaking of our pure energy and exchanging for it his low, impure or dark energy. That perpetrator who was once outside us has not only entered our body through one or more of our orifices but has entered our being, our soul. Those negative influences get introjected inward. You could say that the perpetrator literally takes our light and leaves behind energetic darkness.

So what does acupuncture specifically offer to clean out this unwanted, dark energy and provide us with an opportunity to bring our light, our spark, back into our lives? The process is a point protocol called "Seven Dragons Protocol," which was originally developed by JR Worsley in the tradition of Five Element Acupuncture. Lonny Jarrett, in The Clinical Practice of Chinese Medicine, details the actual point protocols used in the process and provides a thorough explanation for the benefit of Oriental Medicine practitioners. Michael Greenwood, another highly respected and experienced acupuncturist, has also contributed to our understanding of what spirit attachment is and how survivors can be treated for this problem. I think that one of the most important points that can be gleaned from their research, which I have also found to be true in my own exploration into spirit possession, is that acupuncture is not conceiving of or dealing with this phenomenon in the way we might understand spirit possession in the overly dramatized way we see in horror movies. Rather, acupuncture looks upon this negative "spirit" in a subtler way, as something like a holographic overlay to our true being. That overlay is not good for the person who has picked it up, and as acupuncturists we want to do what we can to remove it, to clean it out.

The Seven Dragons Protocol happens to be a valuable protocol to help do that. The treatment is conducted just the same as any of the other approaches we have been discussing—the same fine needles inserted at strategic points in your body.

The Seven Dragons Protocol has two variants, the Internal Dragons Protocol and the External Dragons Protocol. The Internal Dragons Protocol is ideal for someone who has experienced an internal trauma such as rape, and the External Dragons Protocol is more relevant for someone who has experienced trauma from the outside, such as wartime trauma. However, it is still highly useful for rape trauma survivors, especially if they absorbed physical blows during the assault. The goal with these protocols is simply to do everything possible to rid the person of any of those harmful energetic remnants or parasites, whatever they may be or wherever they originated. The reality is that in Oriental Medicine, it's very difficult to facilitate balance and harmony to your body, mind and spirit if you are in any way being impacted by a foreign influence. Taking this step with these two complementary treatments will help your body and being establish a greater receptivity to other Oriental Medicine treatments that your acupuncturist may call upon.

There is no certifiably "one size fits all" acupuncture treatment plan for rape trauma, and, as I mentioned earlier, all acupuncturists will bring their own experience and ideas into how to effectively work with anyone who tells them they have survived a sexual assault or who exhibits many of the symptoms and conditions common to this traumatic experience. From my own perspective, however, the Internal and External Dragons treatments represent a solid part of the foundation of the healing treatment for a survivor. If you are working with an acupuncturist, please share this info with him or her.

© Can Stock Photo / flankerd

Internal Dragons Protocol
(for trauma coming into the interior of the body)

Point | **Treatment Protocol**

CV 14.5 Insert the needles straight, top to bottom, right, then left.

Make sure each needles grab the Qi.

ST 25 Turn all needles in same order 180° counterclockwise.

ST 32 Leave in dispersion for 15 minutes.

ST 41 Check the pulses and the eyes. If there is no change, starting at CV 14.5, pull all needles and feel for a grip. If there is none, then needle location was inaccurate.

Reinsert needles, assuring they grab the Qi, and try again.

If still no change, tonify all needles by turning them 180° clockwise, top to bottom, left, then right.

Leave 3-4 minutes, then pull the needles in reverse order of insertion and seal the points.

If there is still no change, do the External Dragons.

External Dragons Protocol
(for trauma coming from the outside)

Point | **Treatment Protocol**

GV 20 Insert all the needles straight, top to bottom, right, then left.

BL 10 Make sure De Qi is achieved.

BL 23 Turn all needles in same order 180° counterclockwise.

BL 61 Leave in dispersion for 15 minutes.

Check the pulse and eyes. If no change, starting at GV 20, pull all needles and feel for a grip. If there is none, then needle location was inaccurate.

Reinsert needles, get De Qi, and try again.

If still no change, tonify all the needles by turning needles 180° clockwise, top to bottom, left, then right.

Leave 3-4 minutes, then pull the needles in reverse order of insertion and seal the points.

I consider the Seven Dragons treatment EXTREMELY important for many rape survivors. In fact, Five-Element practitioners will say that if other acupuncture treatments are not working, it is because the Evil Qi has not been expelled.

© Can Stock Photo / flankerd

The Ghost Points of Sun Si Miao may be indicated in situations where a trauma continues to haunt an individual long after an event, and the trauma is very difficult to overcome. These point protocols were developed during a time in history where it was seen that illness was often the result of bad spirits. Today we know that this is not the only cause of illness, however modern medicine has practically discarded this possibility and rather categorizes such conditions as psychiatric disorders. We must revisit ancient knowledge and acknowledge that yes, there are some conditions that are very much the result of harmful spiritual energy. It is wise that if one treatment paradigm doesn't work, to try another. The points within the set are not all needled in one session and in each session a different set of 3 points would be needled. The points are grouped into trinities and the groupings represent how deeply a harmful etheric or spiritual energy (Gui) has penetrated into the person's being. The first treatment addresses the most superficial level of invasion, progressing to the last, for the deepest level.

Ghost Point Protocols

FIRST TRINITY – these points are the beginning of the path of psychological dysfunction. There is a buildup of polluted etheric energy (Gui) obstructing the energy of the heart and leading to a disordered spirit (Shen).

1. DU 26 (Ghost Palace)
2. LU 11 (Ghost Faith)
3. SP 1 (Eye of the Ghost).

SECOND TRINITY – Irregular or intermittent flow of Qi. Disordered emotions. The person has no direction in life.

1. PC 7 (Ghost Heart)
2. UB 62 (Ghost Path)
3. DU 16 (Ghost Pillow)

THIRD TRINITY - Alcoholics and addicts. They are stuck here using the alcohol and drugs to get their internal fire going.

1. ST 7 (Ghost Bed)
2. REN 24 (Ghost Market)
3. PC 8 (Ghost Cave)

FOURTH TRINITY – The Gui (polluted etheric energy) has fully taken over and the patient's health deteriorates.

1. DU 23 (Ghost Hall)
2. REN 1 (Ghost Hidden or Ghost Store)
3. LI 11 (Ghost Official)

THE 13th POINT

Gui Feng (Ghost Seal) – This point is located under the tongue at the center of the lingual frenulum. It assists with enlightenment. Also used for schizophrenia or shattered personality. It is not needled but rather pricked and bled.

"The openness and surrender during sexual intercourse can allow the exchange of attached entities between two people. The thoughts, desires and behaviors of an attached entity are experienced as the person's own thoughts, desires and behaviors. The thoughts, feelings, habits and desires do not seem foreign if they have been present for a long time, even from childhood. This is a major factor in the widespread denial of the concept and lack of acceptance of the phenomena of discarnate interference and spirit attachment, obsession or possession."

– Dr. William Baldwin
Spirit Releasement Therapy

I would also include the treatment aimed at clearing the obstruction of the Chong Mai channel in an overall plan because of that vital link of the uterus-heart connection involved. (SP4, P6 point combination). When those points are needled, the patient should feel a very subtle opening of an energy pathway that has been blocking the energy of the uterus from ascending upwards towards the heart. It is very healing and a critical pathway to open after vaginal sexual assault. Opening this blocked energy pathway will help allow the body to heal itself. I also recommend opening the Conception vessel, also known as the REN meridian, using K6, LU7, and in the case of anal assault the Governor vessel or DU meridian using SI3 and BL62.

I also happen to like to bring in a protocol called "Buddha's Triangle," which is good at deeply nourishing the spirit, the acupuncture points H7 (Heart 7) nourishes the heart transforming anxiety and worry, P6 (Pericardium 6) opens the energy of the chest and lifts the spirit and the LU9 (Lung 9) will alleviate overall feeling of sadness, depression, heaviness of the heart. You can always add DU20 at the vertex of the head for good measure to raise energy in the body to counteract the sinking feeling that sadness or depression can cause.

Another treatment that can fit an acupuncture plan for treating rape trauma survivors is the Yin/Yang Tuning Fork Pattern developed by Mikio Sankey, who focuses on Esoteric Acupuncture. I find this procedure helpful because of the critical need to restore the balance of yin and yang. I'll add one another approach to this personal list of treatments that can be especially helpful for survivors. Of course, there are many others that will naturally enter the picture based on the assessment of any individual seeking Oriental Medicine treatment for the pain and suffering linked to rape trauma. This other approach is called "Seitai Shinpo," a Japanese modality that I trained in. This treatment can be highly

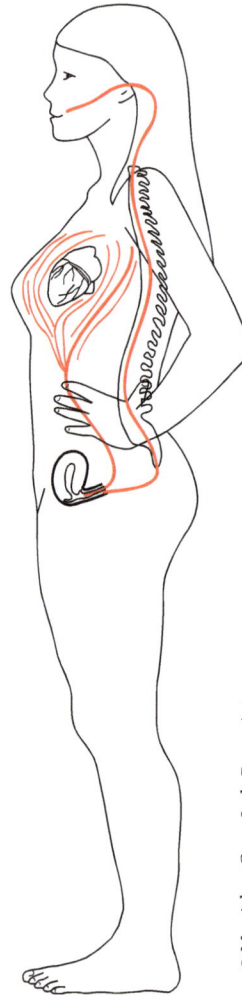

Pathway of the Chong Mai energy meridian

© Morrighan Cox – Lady Raven Art

effective for survivors, especially those whose trauma is old and has resulted in seemingly calcified negative emotional states. Seitai Shinpo moves the Qi and the blood at a very deep level. Interestingly, the protocol shares many of the acupoints used in the Draining Aggressive Energy treatment. As mentioned earlier in this book, emotions are carried in the blood in Oriental Medicine. Thus, emotional energy that has been stuck for a long time will typically negatively shape the constitution and personality of the person. Seitai Shinpo helps to invigorate and harmonize the organs, which besides their commonly known functions also govern various emotions. Anyone experiencing this treatment will notice that a different technique is involved, which consists of what acupuncturists call dispersion needling. That means that stuck energies are mobilized by needling relatively deeply into a targeted point, then we remove the needle and needle into the next point. The treatment is then reinforced with what is called direct moxibustion, an old Japanese technique where a tiny pinch of an herb (about the size of a sesame seed) is placed on an ointment barrier on top of an acupoint that has been needled. The moxa—commonly known as mugwort—has a deeply penetrating action, dislodging stuck energy (and yes, this includes deep, stuck emotional energy) deep in the body. There are a limited number of acupuncturists trained in this highly effective technique, but if you are limited in your resources and are unable to travel to a Seitai Shinpo acupuncturist, you can substitute a treatment that simply strongly moves Qi and blood (4 gates - LV3, LI4 - UB16 and other points your acupuncturist might select) and follow it with a Draining Aggressive Energy treatment.

Some years ago, I was treating a woman who had come to my office seeking treatment for an issue that had no apparent connection to sexual assault. As we were completing the Seitai Shinpo treatment, she had a major release of tears. This woman, who was in her fifties, told me that she had just gone through an amazing experience on my treatment table. She had been sexually abused as a young girl, which was not a part of her history that she shared during out initial intake process. Even after years of therapeutic treatment directed at healing from that trauma, she knew that she still carried remnants of the experience with her. And those pieces felt like they were being dissolved on the table with those cathartic tears. Seitai Shinpo had been successful in accessing a traumatic event that old, and that deep! For adults who have childhood sexual abuse issues that may have shaped their emotional constitution in a negative way, over the long term, I highly recommend the deep healing effect that Seitai Shinpo can offer.

In treating a rape trauma survivor, the Seitai Shinpo treatment could be followed up with a second treatment of Draining Aggressive Energy, though this would be a very long treatment and the patient and practitioner should be prepared for it. The positive effect of utilizing both of these treatments can be compared to what happens when you're trying to launder a really dirty pair of jeans in the washing machine. When you put those jeans in the washer with your soap, you agitate, agitate and agitate. That's just what Seitai Shinpo does— energetically it stirs things up from a deep level so it can be harmonized. Just as a washing machine agitates clothes then rinses them, the Seitai Shinpo treatment is the agitation part of the cycle and the Aggressive Energy Drain is the rinse part of the cycle.

Now, whichever treatments your acupuncturist utilizes in assisting you on your healing journey, it's important for you to follow up with the Chinese herbs that she or he recommends.

Yunnan Baiyao is an herbal formula that slows down blood flow, thereby slowing down what may be intense and uncontrolled emotional expression. *Yunnan Baiyao* was originally designed to astringe (pull in) blood after a gunshot wound. Since emotions are energy in motion, in the case of uncontrollable emotional expression, this formula can harness and restrain the emotions to make their expression, or release, manageable.

The classical formula *Sheng Mai San*, also known as Generate the Pulse Powder, can address a depressive function of the liver caused by early trauma such as sexual abuse. *Gui Zhi Jia Long Gu Mu Li Tang* can be utilized for harmonizing the ying, our internal nutritive energy—our innermost nourishing energy that we share during a sexual union—and our Wei Qi, the outermost energetic boundary of our subtle body that's so important in our protection from outside forces.

Chinese herbs can have a profound, positive effect in restoring our balance and harmony, and they work well with acupuncture. However, it's important to call upon them only as directed by someone well trained in Oriental Medicine, to avoid any potential misuse.

Moxa, the warming herb that is used in the moxibustion technique, can be used in other ways. If your acupuncturist approves of this technique, here's something you may be able to do at home on your own. If you feel empty and cold in your pelvic area, in a way that leaves you feeling generally unhappy and depleted, you can actually "warm the womb" to replenish depleted energy. The process involves lighting a special type of charcoal moxa cone, also known as Jwahoon moxa, with a lighter and then placing the moxa cone on a pan on the floor. Then, you sit on a short stool with a hole in it or squat over the lit moxa.

The heat from the charcoal moxa creates a warm, nourishing sensation in your womb.

It is also possible to use a moxa pole, which is a cigar shaped roll of mugwort (moxa) herb and other herbs in a paper wrapper to warm the point 1.5" below the navel. This will build Qi or life force energy, particularly in the lower part of the body. This will combat a sense of depletion—an emptiness that can be rebuilt. Have your acupuncturist demonstrate the technique and teach you how to do this safely at home. It works well to build back energy in the lower abdomen and pelvic area.

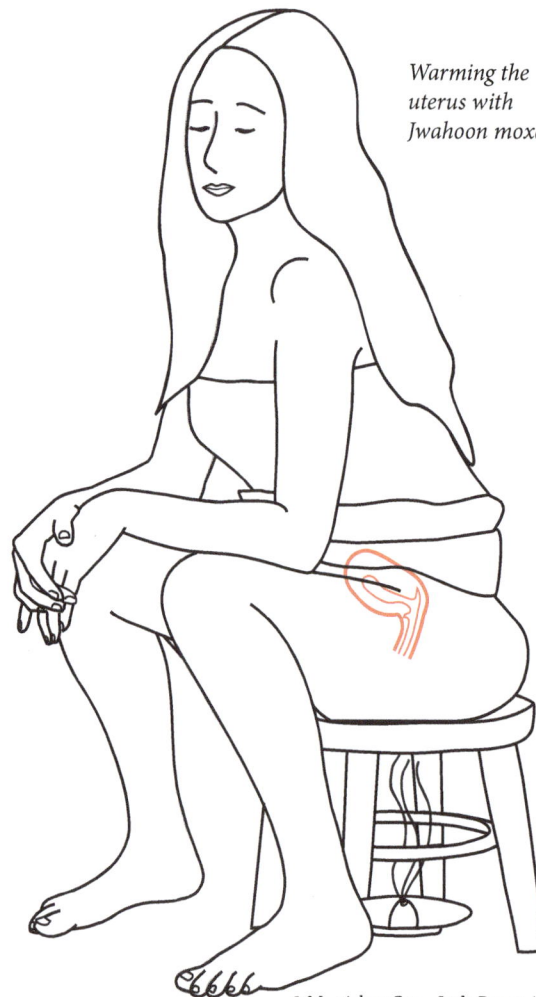

Warming the uterus with Jwahoon moxa.

© Morrighan Cox – Lady Raven Art

In Oriental Medicine, another name for our vital life force sexual energy is "Jing," which we introduced in our discussion of depletion back in Chapter 1. There's now an entire classification of Chinese herbs called "Jing Tonics" that can contribute to restoring that essence. Those herbal formulas, of course, will work best if used in conjunction with acupuncture treatments focused on the same desired outcome.

Everything about Oriental Medicine is about harmonizing and balancing. That's a major part of how this modality is so powerful for treating not only rape trauma but many other problems and conditions.

Acupuncturists are always looking out for ways to not only alleviate suffering but to usher in profound depths of healing and positive energy. Those opportunities often emerge directly after one or more treatments have successfully addressed what was holding a person down. Earlier in this chapter we explored how dark or even evil spiritual energy can enter our body and being through the many orifices of our body. That's how spirit attachment can occur. The flip side of this picture is that those same orifices can be the entry points of sacredness and purity.

Let's say your acupuncturist has been working to free you from intense emotions and other sources of pain over the course of several treatments. He or she has completed the Draining Aggressive Energy and the Seven Dragons treatments, along with other approaches selected specifically for your personal healing process. Now it may be time to focus on approaches with a focus on purification.

Energetically, that divine spark or soul that seems to have been lost ever since you were sexually assaulted may not be so hard to resurrect. In fact, it still may be anchored to your body etherically, just waiting to inhabit your body once again. Your acupuncturist may be able to call upon treatments aimed at ushering that spark back into you. One potential approach would be to treat upper points on the kidney channel. These points are located at the top of the chest, in between the ribs on either side of the sternum. There are specific points where treatment is directed to unblock the expression of the spirit of the heart, resurrect the spirit from a place of emptiness or depletion and remove tarnish from original nature or divine essence. In fact, the names of some of these points are: Spirit Seal (K23), Spirit Ruin (K24), Spirit Storehouse (K25).

Imagine reaching a stage in your healing that restores your true nature, so that your life unfolds the way that you want it to, and not as a response to trauma. That's the kind of image, and hope, that can sustain any of us as survivors as we reach out for healing assistance through Oriental Medicine and other modalities that can reach us at our mind/body/spirit core.

© Can Stock Photo / tanais

Summary of suggested therapeutic interventions

The specific acupuncture protocols listed in this book are meant to be shared with an acupuncturist. It is always up to the practitioner to decide what treatment is warranted after conducting a thorough intake.

Chapter 5
Expanding Your Healing Options:

Other Body-Oriented Therapies

As we have learned in the last two chapters, acupuncture offers a powerful opportunity for anyone seeking relief from sexual trauma. Oriental medicine provides a wide spectrum of approaches and treatments that address the multiple-level effects of trauma from rape, sexual assault and sexual abuse. Working with an acupuncturist who understands your vital needs and has the experience to facilitate this treatment can open a wide door to your healing in body, mind and spirit, especially when combined with conventional talk therapy or complementary therapies that we will discuss next.

As a Licensed Acupuncturist, I always feel encouraged whenever I hear that more sexual trauma survivors have become aware of what Oriental medicine can do to bring their lives into greater balance and harmony. And as a survivor deeply committed to expanding the awareness of all possible avenues for healing available to those suffering from sexual trauma, I'm also excited to share what I have learned about other body-oriented therapies and approaches that could begin to alleviate their pain and suffering. As I mentioned in the Introduction to this book, there is a wealth of resources out there for survivors these days, including many modalities that you may not yet have heard of, and I want you to have every possible option at your fingertips when you launch or continue your personal healing journey.

So in this chapter we're going to expand your healing options by taking a brief look at many of these effective treatment options. We can't cover every type of treatment that could potentially help heal the effects of sexual trauma, but we can explore a wide enough sampling to offer more choices along your healing path. I am not professionally trained in these other treatment approaches, so I am simply passing on what I have learned in my own research, my personal experience or through circles of body-oriented practitioners.

Many of these other body-oriented therapies were developed with a foundation that has much in common with the beliefs and practices of Oriental medicine. Some of them incorporate the same energy points used in acupuncture. These modalities include:

Eye Movement Desensitization and Reprocessing (EMDR)

EMDR, which was developed by psychologist Francine Shapiro, can be an especially effective therapeutic tool for rape trauma survivors because it involves recalling and re-experiencing the effects of a traumatic event with the goal of transforming the thoughts, feelings and behavior previously associated with the negative experience. EMDR is built on the premise that the mind can heal from psychological trauma just as the body can heal from physical trauma. It has been so successful in healing PTSD that it has gained the recommendation of the Department of Veterans Affairs and the Department of Defense.

Therapists trained in this technique engage clients in an experiential process that may include rapid eye movements, the shining of a light or the tapping of a wand or a finger on the client's leg while guiding them through the traumatic event and the accompanying body sensations and emotions. Successful EMDR treatments result in reducing or eliminating the trauma and substituting a positive belief for the negative experience to empower the person to navigate life in a healthier way. The network of EMDR-trained therapists has been growing rapidly in recent years, with many mainstream therapists calling upon this technique as an adjunct therapy.

Emotional Freedom Technique (EFT™)

Like acupuncture, EFT™ works with the meridian system of the body. It also aims to get energy moving so the body can heal itself. This modality, which was developed by Gary Craig, basically involves invoking a traumatic memory, then tapping with your fingertips on several specific acupoints repeating affirmations. The tapping sends signals directly to those parts of the brain where emotions are stored so that positive changes may begin to occur.

EFT™, like acupuncture and EMDR, also has a track record of success treating veterans of war with PTSD, so it certainly can assist sexual assault survivors with their form of PTSD symptoms. This tapping approach is something you can do with the guidance of someone with experience utilizing the system, but it's also something you can learn to do on your own. This can be especially attractive to survivors with limited financial resources. You can also track the results immediately because the process entails naming a troubling issue that you want to work on, rating the severity of it, proceeding through the specific steps of the tapping process while repeating a specific affirmation statement, and then retesting the severity of the issue.

Many practitioners and proponents of EFT™ now offer their own presentations of the technique online. If you're interested in the possibility of integrating this healing approach in your path, you'll have no trouble finding them.

TAT® (Tapas Acupressure Technique®)

The connection to acupuncture is especially clear with the approach of TAT® because it was developed by Tapas Fleming, a Licensed Acupuncturist and experienced practitioner of Traditional Chinese Medicine. As she explains the genesis of the technique, the idea came to her while napping in her office one day.

This is another technique you can learn to do on your own or work with someone trained in the modality. It involves placing your hands on specific acupoints on your head and face while moving through a series of statements linked to a problem that you want to address. The goal is to free parts of yourself stuck in a trauma or negative belief and become present in the here and now. So this is another technique that fits naturally with the issues and symptoms related to sexual trauma. The foundation of this approach is geared to bring balance and healing on the physical, mental, emotional and spiritual levels. In one of her online videos, Tapas Fleming specifically addresses how TAT® can help you deal with what we have been describing as that dark energy that stays with you after sexual assault and limits you in many dimensions of your life until you find a way to release it.

The TARA Approach

The TARA Approach for the Resolution of Shock and Trauma integrates the usage of Jin Shin Tara, a subtle energy system based on the Extraordinary Meridians, and targeted language sensitive to the nervous system as well as aspects of Cranio Sacral therapy. Dr. Stephanie Mines, a neuropsychologist, created this technique, which you also can learn to use independently. It has been taught in hospitals, universities and social service agencies, as well as domestic violence shelters and midwifery schools.

Since it is a form of subtle energy medicine, The TARA Approach also is well positioned to treat the often-invisible injury of sexual trauma. This particular treatment approach aims to distinguish between trauma and shock, which is described as the most severe form of trauma and a cause of long-term chronic conditions deeply rooted in the body. Sexual assault and child sexual abuse certainly can trigger the kind of shock outlined in this context.

Zero Balancing

Fritz Smith, MD, developed this body-mind therapy back in the early 1970s. A trained Zero Balancing practitioner uses touch to address the connection between parts of the body and energy. By using finger pressure and gentle traction on targeted areas of pain or tension on the body, the practitioner creates what are called points of balance, which become the center for the reorganizing and relaxing of the body. In a typical session, the person receiving Zero Balancing treatment will experience focused attention on both the deep and dense tissues of the body along with soft tissue and energy fields.

The goal in this approach is similar to the goal of acupuncture: to clear blocks in your body's energy flow to usher in a feeling of balance and harmony. It seeks to increase your life energy so that you can be more of who you are, free of unwanted pain and suffering.

Jin Shin Jyutsu®

Jin Shin Jyutsu® is an art of harmonizing the life energy in the body. It was developed in Japan in the early 1900s by Master Jiro Murai, but it wasn't introduced in the West until one of his students, Mary Burmeister, began sharing it some fifty years later. It is built on the familiar premise that for the body to be healthy, energy needs to flow freely. When energy is blocked, disharmony results.

This is the understanding on which Oriental Medicine is built, so anyone who works with a Jin Shin Jyutsu practitioner seeking relief from the symptoms of sexual trauma can feel confident that a similar approach will be followed. The practitioner uses his or her hands on specific locations (known as "Safety Energy Locks") in the body that are central to that free flow of energy. Like acupuncture, the success of Jin Shin Jyutsu treatment is steered not only by the effective usage of the art but by the spirit in which the work is approached. Uniquely, Jin Shin Jyutsu offers a rich body of Self Help material, which is the foundation of the art, and which can be easily and effectively applied any time of day. The whole philosophy of this art is to encourage a life of harmony and Self empowerment, in contrast to the chaos and disturbance of a life in the grasp of sexual trauma that has not yet been effectively treated.

For further information, please go to www.JSJinc.net.

Acutonics®

This healing methodology is rooted in classical Chinese medicine, psychology, science and sound. Tuning forks are applied to acupuncture points, chakras and points of pain. Acutonics® may be used in conjunction with acupuncture needles or as a stand-alone therapeutic approach to the treatment of trauma. The idea is the same: circulate energy in the meridian pathways to clear out blocks and get the energy moving freely again. This technique has been especially useful in treating anyone with a fear of needles as well as young children and critically ill patients.

By working with two tuning forks—which when combined together create a musical interval—a specific vibratory energetic with its own unique healing properties is created. When these frequencies are applied directly to the acupuncture points, it is possible to access the body's core energetic system in a noninvasive way to effectively move, nourish and balance the Qi. Some sound intervals are excellent in promoting a sense of balance and reducing anxiety,

63

others lift the spirit and promote a feeling of joy, while others move or break up stagnation.

Of particular relevance to the treatment of sexual trauma, trans-generational trauma and all types of traumatic stress is the focus within Acutonics® of working with the Eight Extraordinary Vessels. The vessels are a pre-meridian network formed from the initial cell division, a matrix that harmonizes all physiological systems and are especially responsive to and conductive of sound vibration. They reconnect us to Wu Chi, to the One, and are the true roots and reservoirs of all the body's energetic and physiological forms that draw from the infinite seas of heaven and help bring the soul to a place of remembrance of wholeness.

The Acutonics® System was initially developed and tested beginning in 1995 at an acupuncture school in Seattle, Washington by Donna Carey, who was Clinical Dean, and Marjorie de Muynck, who was a musician and shiatsu instructor. From those early years under the direction of Ellen F. Franklin and Donna Carey it has grown into the Acutonics® Institute of Integrative Medicine, LLC, which is celebrating its 21st birthday. Acutonics® offers an in-depth certification program in the applied use of sound vibration in a clinical setting. There are more than thirty certified teachers who offer Acutonics® programs around the world. For more information visit www.acutonics.com.

Sound Healing

Sound can play many roles in the healing of survivors of sexual trauma. I mentioned earlier that you may benefit from playing gentle, natural music at home during sleep between acupuncture treatments. Some practitioners of Oriental Medicine even incorporate sound in their treatments.

There's a body of knowledge to draw from in this kind of integration called Cymatics. Hans Jenny, a Swiss medical doctor and natural scientist, conducted a pioneering study of wave phenomena in which he demonstrated how sound can manifest itself in various materials, many in beautiful kaleidoscope images. One popular experiment shows how to sprinkle sand on a metal plate and then vibrate the plate with musical sounds, causing the sand to be shaped in forms similar to a flower or mandala or snowflake that change from one to another. Many healing practitioners apply the same principles to transmit sound waves through the body to restore harmony in the body's energy flow. This is especially relevant for the healthy unfolding of a woman's sexual energy! Instead of her energy being knotted or rigid due to trauma, sound frequencies can help shake up and harmonize that energy and bring back a healthy flow.

In acupuncture, this technique can be used after a treatment such as opening the Chong Meridian energy pathway. A sound healer can play a digeridoo or a wooden flute directed toward the pelvic area or another affected area to further support the work being done to encourage healing. One acupuncturist I knew had a treatment table built with harp strings underneath it and he would actually play the harp strings while his patient was lying comfortably with the needles inserted. In this way, he was able to help a patient heal from a chronic lung condition that the allopathic doctors said was terminal. I believe that there's almost no limit to the ways in which sound can help sexual trauma or anyone on a healing journey! In fact, many country songs are musical expressions of healing of betrayal or heartbreak. To quote the famous country singer Reba McEntire: For me, singing sad songs often has a way of healing a situation. It gets the hurt out in the open into the light, out of the darkness. Don't underestimate the power of music or sound.

Hara Line Repair

The Hara Lines are an aspect of our toroidal field. Toroidal fields are how energy flows in all things. It's how energy moves in a closed system such as cells, people, the earth, the universe.

While the Hara Lines run through the center of the body from above the head to below the feet, they are often more easily psychically perceived in the area of the abdomen in their color, determined by their frequency. The structure is 'rope-like', with the abdomen area being where the smaller strands separate from the Main Hara lines 1 and 2 to create lines 3,4,5,6 and 7. Rita Marr, who was fortunate to be given new and unique channeled information about the Hara aspect of the human energy field, calls the seven Hara Lines The Seven Keys to Health, Harmony, Success and Abundance.

Rita Marr: www.theharaline.com

The Hara Line is also known as the Tai Ji or Tai Chi pole, Central Axis, Central Pillar or Main Central Flow. It is the central energetic axis of our body. Hara Line repair is critical in the healing of rape trauma as this energy pathway is often broken or blocked. Opening the Chong Meridian (GB41 + P6 points), then strengthening the axis using DU20 (crown), CV17 (heart) and K1 (grounding) with the assistance of an acupuncturist helps tremendously. The more energy you can run through your Hara Line, the stronger you will be.

Barbara Brennan has specific Hara Line techniques that can be researched online.

Angela Shelton is an activist in the area of sexual assault. At an event, she had survivors perform a very effective Hara Line repair technique. The technique was simple. She placed cushions onto a chair and, one by one, had survivors strike the cushions with a baseball bat. The key here is to get energy flowing through the entire body, from feet through torso to arms. The strong current of energy moving fully through the body can dislodge stuck energy along the Hara Line. The survivor can also integrate their voice, by yelling from a deep place, allowing the voice to resonate through the whole body.

Other ways to unblock the Hara Line include using a few rounds of the Sun Salutation yoga pose series, which moves and aligns energy in the core, followed by vocalizing from a deep place. Imagine humming loudly or chanting from the lowest note you can sing to the highest, allowing the sound to resonate through the core of your entire body. The area where the notes are off is where there is damage along the Hara Line. It is possible, with practice, to vocally work out this energy kink and open the Hara Line. In martial arts, there is a vocalization technique called the Kiai. This is a powerful vocalization, also known as the Spirit Yell, a type of battle cry. When done correctly, it is said that you can almost hear your ancestors coming through you. We truly are multidimensional interconnected beings! Also, there are a multitude of healers that you can find online who specialize in Hara Line repair.

Here is a gentle way to harmonize energy in the central channel, from the Jin Shin Jyutsu website: "The Artless Art of Getting to Know (Help) Myself"

The Main Central Vertical Flow is the source of our life energy. This pathway flows down the center of the front of the body and up the back of the body. Here is a Jin Shin Jyutsu® self-help process to harmonize this pathway.

Harmonizing the Main Central regularly helps you feel centered and ensures that you will have plenty of energy. Some people find it calming and use it to fall asleep, while others like to use it to clear away the cobwebs upon awakening. For optimum results, do this daily.

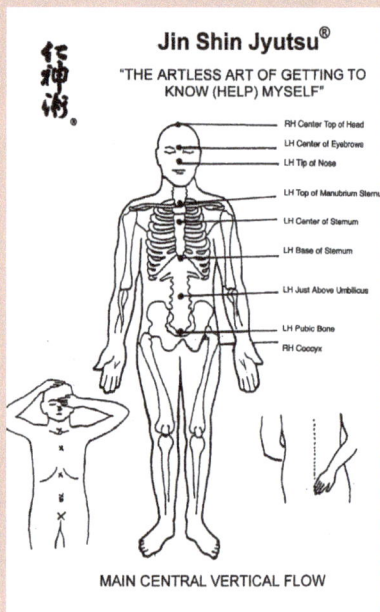

Jin Shin Jyutsu®
"THE ARTLESS ART OF GETTING TO KNOW (HELP) MYSELF"

RH Center Top of Head
LH Center of Eyebrows
LH Tip of Nose
LH Top of Manubrium Sternu
LH Center of Sternum
LH Base of Sternum
LH Just Above Umbilicus
LH Pubic Bone
RH Coccyx

MAIN CENTRAL VERTICAL FLOW

Main Central Vertical Flow

Step 1: Place the fingers of the right hand on the top of the head (where they will remain until step 6). Place the fingers of the left hand on your forehead between your eyebrows. Hold for 2 to 5 minutes or until the pulses you feel at your fingertips synchronize with each other.

Step 2: Now move the left fingertips to the tip of the nose. Hold them there for 2 to 5 minutes, or until the pulses synchronize.

Step 3: Move the left fingertips to your sternum (center of your chest between your breasts). Stay there for 2 to 5 minutes or until the pulses synchronize.

Step 4: Move your fingers to the base of your sternum (center of where your ribs start, above the stomach). Hold them there for 2 to 5 minutes, or until the pulses synchronize.

Step 5: Move your fingers to the top of your pubic bone (above the genitals, center). Stay there for 2 to 5 minutes, or until the pulses synchronize.

Step 6: Keep your left fingertips in place and move your right fingertips to cover your coccyx (tailbone). Hold for 2 to 5 minutes or until the pulses you feel at your fingertips synchronize with each other.

Note: The right hand remains on the top of the head while the left hand moves down the body until the final step.

Movement and Bodywork

There's a wide range of movement practices and bodywork techniques that survivors of sexual trauma may choose to explore. Any time you get your body moving, you have the opportunity to at least begin to encourage the release of blocks of energy in your body. Unfortunately, many survivors caught in a web of depression and hopelessness resist the idea of moving their bodies. It may sound easier to simply pull up the covers and try to hide from the painful feelings and protect yourself from the world.

At some point, however, you realize that's not a workable or permanent solution and simply waiting for time to heal wounds is not effective. We all have a natural desire to heal. Once that desire is tapped, you decide that you're ready to seek help through various forms of therapy and treatment, which may include acupuncture and the body-oriented therapies with close links to Oriental Medicine that we have mentioned so far in this chapter. When you begin to look around at the options of what can help you address the deep pain and suffering you are experiencing, you may also want to keep in mind other movement and bodywork techniques that have the potential of boosting your recovery and healing from sexual trauma.

Some of these approaches may already be familiar to you, although you may not have typically looked upon them as offering assistance for your needs related to recovering from sexual assault. Others may be outside your personal sphere of awareness now, but as you pursue your healing you find that expanding that awareness is a natural way to enhance the possibilities of making positive change in your life. We'll start with one form of movement that almost everyone has heard of these days:

Yoga and the Potential for Healing

In the last couple of decades, yoga has become a popular resource for fitness and relaxation for millions of people in the West. Most people understand the benefits of yoga in reducing stress and enhancing flexibility. What many people don't realize, however, is that yoga can also be called upon to summon a greater capacity for healing—including healing from sexual trauma.

As an acupuncturist, at times I have jokingly referred to acupuncture as "yoga for lazy people." The reason is that if you practice yoga every day, you get your Qi energy moving and you align and balance the flow of that energy in the body. With sexual trauma that lives on in the body long after the traumatic event, getting that Qi flowing and moving in the right direction is central to your healing. Yoga can't do everything that acupuncture can do, but it can certainly guide you on an important step toward where you want to go.

Many yoga instructors have come to recognize the healing potential of this popular movement practice, so it's not surprising that a connection has been made between what yoga offers and what survivors of sexual trauma need. Zabie Yamasaki, a yoga instructor and activist who is a survivor of sexual trauma herself, brought this connection to life as the founder of Transcending Sexual Trauma through Yoga. This organization offers group and individual yoga classes for survivors.

The program helps survivors address both the physical and emotional pain caused by sexual assault. Through a directed yoga practice, along with complementary exercises that may include art therapy and journaling, survivors are steered toward reclaiming their bodies and building a stronger connection to self and others.

Yoga and Healing from Trauma

Yoga was always one of the main healing modalities on my long journey to mental and physical health as a survivor of child sex trafficking and extreme violence, sold in 1969 at age six into an international, murderous pedophile network. I was dramatically rescued in at age eleven, when one of my perpetrators made a last-minute deal for my life.

I entered psychotherapy in the eighties, which soon provided a first opening of the psychic wound. I cried for three weeks. With my tears connected to their cause, I experienced the magic of neural integration. Through the grief, I received a taste of what felt like truth, for the first time in my adult life. Getting closer to my true self with each leg of the healing journey, the gifts were worth the pain. I gradually felt into all the devastation and betrayal I had suffered. It is something I would not wish upon anyone.

The physical practice of yoga started in 1994 came as a natural extension of the sports I had done since my rescue as a young teen to ensure that the injuries I sustained in the network would not render me crippled. From the first yoga class, I knew I had found the perfect physical therapy. Aligning the breath with movement was the key. Allowing for the parasympathetic nervous system to engage through a simple lengthened exhale created instant peace.

Before I started practicing yoga, I had no body awareness below the neck. The yoga practice helped me to become present inside my body. This increasing self-awareness also made me self-conscious. I didn't like having anyone behind me, or having teachers touch me without asking permission. Yet, I wasn't healed enough to stand up for my needs. The yoga practice brought both healing and put obstacles in the way of healing.

After reading "Autobiography of a Yogi" by Paramahansa Yogananda, I began to learn the meditation techniques of the Self Realization Fellowship. These teachings offered the largest context for my childhood experiences. The spiritual perspective is like a dome underneath which all my experiences exist beyond judgment, self-blame or pain. Meditation navigates the space between who I think I am and who I am. Without the larger context of spirituality, I would never have been able to heal as well as I have.

The combination of meditation and yoga, with therapy and writing, created the path of healing for me. In 2010, I wanted to share this healing and went to teach yoga into the prisons, which led to the creation of my non-profit organization Liberation Prison Yoga in 2014. As soon as I entered the prison to teach yoga, I felt that I had found my purpose in life. My healing had been a long and lonely road, but once I got to share the tools I had discovered, my heart opened and healing increased exponentially. I never went in to receive anything, but I got so filled up with the genuine sweetness and generosity of the students that I could barely contain the love.

I created programs that would offer what was missing when I was practicing yoga in studios, with discussion, journaling and meditation as well as a physical practice. I left all those things out of the programs that did not work for me in the yoga classes, such as commands in the language, adjustments, mats in rows (we put them in a circle). Liberation Prison Yoga's teachers focus first on connection from the heart.

The work has taught me humility. I have met people who bravely face dire circumstances with more grace than can be imagined. Humility is possible only when one can draw from the wealth within rather than look outside for approval and confirmation. Seeking humility in the context of yoga is important to all who practice, to move away from the excessive focus on the physical and towards service and the potential of yoga as a complete path to enlightenment.

Anneke Lucas
Child Sex Trafficking Survivor
Founder of Liberation Prison Yoga, TED speaker

Tension & Trauma Release Exercises (TRE®)

Do you remember in Chapter 1 when I mentioned that there is a body-oriented technique that helps people to shake off the effects of trauma in a way similar to what animals in the wild do? TRE® is that technique. Created by Dr. David Berceli, Ph.D., the Tension & Trauma Release Exercises are based on the understanding that tension and trauma are both physical and psychological or emotional. The process unfolds through a series of exercises designed to assist the body in releasing the deep muscular patterns of tension and trauma.

So how does shaking enter the picture? The exercises start off by building up tension in specific muscle groups and conclude with one exercise in which you are lying on the floor. The final posture is designed to induce a natural reflex mechanism of shaking. It is the act of shaking or vibrating that releases the tension and emotional trauma while promoting a deeper relaxation and increased resiliency in the body. From my perspective as an acupuncturist, I would say that the shaking, which can last for several minutes, helps to bring back the pulsation of our life force in the areas of our body where the energy had become stagnant and the musculature was holding on to that trauma. You can learn about the theory behind this approach by reading any of the following books that make the direct connection between this kind of induced shaking and what prey animals do after being traumatized by a predator. A few books that describe this process are:

Shake it off Naturally by David Berceli
Why Zebras Don't Get Ulcers by Robert Sapolsky
Walking the Tiger: Healing Trauma by Peter Levine

TRE® has been taught by trained practitioners to large groups, including military personnel. Once you learn the process, you are empowered to continue using the shaking practice on your own. Again, this is a practice that you can try without having to "talk out" the memories and associations linked to sexual trauma.

Reichian Therapy

As we have learned, survivors of sexual trauma hold memories and emotions in a frozen or blocked state in their bodies, and those frozen pieces need to be released for healing to occur. Reichian Therapy seeks to do that through an integrated approach that brings together deep tissue massage, guided breathing and dream analysis or other more psychological talking tools. Wilhelm Reich, who once studied with Freud, developed this technique and its philosophical foundation that people tend to build walls of armoring that they believe will protect them from the world but that actually have the effect of crippling their capacity to experience a full and joyful life. Reichian Therapy seeks to break down that armoring and enhance one's capacity to live and love more freely.

What's helpful about this approach for survivors is that the practitioner seeks to locate the specific parts of our body where emotions are stored. Is it the chest, the pelvic area or somewhere else? Once they pinpoint the impacted areas that have triggered the armoring, they focus their manipulations on that area to get the energy moving again—to get the golf ball of congested emotions unstuck from the garden hose so the emotional river can freely flow. It's also worth noting that Reichian Therapy initially evolved from the Freudian view that all neurosis is caused by sexual frustration or repressed sexuality. Over the years, most practitioners of this technique recognized that a wide variety of triggers induce suffering and trauma, and healing encompasses all dimensions of life. However, the foundational understanding of blocked sexual energy can be especially relevant for survivors who seek to bring back that vital component to their lives.

Polarity Therapy

Polarity Therapy is another useful technique that manipulates the body to release tension and balance life energy. This bodywork approach was developed by Dr. Randolph Stone, a chiropractor, naturopath and osteopath. A Polarity Therapy practitioner may integrate nutrition, exercise and counseling along with the bodywork. Like an acupuncturist, a trained Polarity therapist listens and observes to sense which areas of the body need to be touched to free up the Qi or life energy.

The touch utilized by the practitioner may by a deeply penetrating approach, or it may take the form of a molded contact and gentle rocking. Throughout the treatment, the therapist is seeking to release tension from those long-held patterns and get the energy moving again. Those who receive this treatment sometimes experience an emotional release of tears or even laughter during a treatment.

With these and other bodywork approaches, it's important that survivors feel comfortable with the process before committing to an initial session. Just as you would in considering working with an acupuncturist, you will want to take your time selecting a practitioner and asking any and all questions that will help you understand how the technique works, what the treatment goals are, and what you can specifically expect during treatment.

Sometimes you might try a new approach once and decide that it's not a fit for you, which is fine. But you also have the reassuring experience that you have found another ally in your mission to chip away at your sexual trauma and eventually usher you into a more balanced and fulfilling life. Whether you consider these options with an I'll-try-anything-once mindset or a far more cautious attitude that aims to narrow down the field of

candidates to aid and assist you in your healing, your intuition will help steer you toward the choices that will serve you.

Tai chi, Qigong

Some survivors find they can benefit from learning and practicing the simple Chinese exercises of Tai chi and Qigong. Those who have suffered an experience such as sexual abuse, particularly from childhood, may have very little idea of how energy naturally feels moving through their bodies. These ancient Chinese exercises, which have become more commonly taught in classes in the West in recent years, help to activate the healthy flow of energy through the different meridian systems.

Tai chi trains the Qi to flow through our body in a slow, healthy, balanced way. The slow, rhythmic movements of Qigong, which also includes breathing and mediation, open our channels and encourage us to move gently through our day at a speed supportive of life. Once you learn these exercises, which usually take no more than 15 or 20 minutes at a time, you can continue to practice them on your own. It's one more way to bring you closer to the balanced and more peaceful life that you deserve.

Massage and Other Nurturing Tools

No matter how long it has been since you were sexually assaulted, your body and being can always benefit from processes that bring a nurturing quality. This is especially welcome when the nurturance is delivered in the areas of your body that suffered the greatest impact.

Mayan Uterine Massage, a technique that any trained massage therapist can easily learn, is a physical manipulation of the uterus through the lower abdomen. The practitioner trained in this technique would use directed strokes and pressure to enhance your blood circulation and tone the uterus. The technique can free up a great deal of stored emotions in the uterus. This ancient form of massage is also used to nurture the uterus after childbirth, as well as specific conditions such as uterine fibroids, but it can also deliver soothing comfort to women who have survived sexual trauma.

Karsai Nei Tsang, which is also referred to as therapeutic massage for the sexual organs, is a technique that I experienced myself at Qigong master Mantak Chia's retreat in Thailand. While the process is directed toward releasing blockages in the groin area, the procedure itself works more with the channels that go into that region of the body. This would primarily be the Liver channel from Oriental Medicine, which starts at the inside of the big toe and travels up the leg, through the groin and up the side of the torso. The body manipulation is focused more on the abdomen and lower abdomen, but there is a small amount of vaginal penetration using the gloved hand of the practitioner. This may be surprising or even seem inappropriate to Western minded patients, but if informed consent between the practitioner and patient has been obtained, it is not any more "strange" than a routine gynecological exam, since the technique is administered in a healing space with the correct intention. Moreover, the trained practitioners of this modality are required to clear their own traumas so that their own energy is balanced and clear before working with patients. The Karsai Nei Tsang massage technique can be very effective in bringing back that pulsation of life force that can become congested in the groin area because of repeated sexual trauma. The technique is particularly effective for conditions of armoring. Because of the nature of this therapy, I would only recommend it in the later stages of healing, after trauma has been cleared and resolved and deeper body work is desired. Since most readers won't be able travel to Thailand for this treatment, it is possible to ask a trusted local massage therapist to learn the technique though a book or DVD that teaches the technique

(see Bibliography) and it can even be modified for the patient's comfort level.

Triple Goddess Rose Oil Vaginal Serum is a product that you can purchase from snowlotus.org and apply yourself. This soothing, nourishing serum developed by Licensed Acupuncturist Rachel Koenig contains pure Rose essential oil in an organic carrier oil and can be used topically and internally. It's an extremely high frequency essential oil, and the fragrance itself carries a nurturing quality. It also can be helpful just to remind yourself of how the rose is so prominently associated with blossoming and blooming, which can be a powerful antidote when you feel that energy is stagnant in the groin area. The serum can be used overnight in conjunction with a rose quartz yoni egg to embody the healing properties of not only the essential oil of rose but the rose quartz gemstone. Please research the correct way to use yoni eggs before you use them.

Vaginal herbal steaming is an age-old natural remedy that many women have called upon to cleanse the vagina and uterus for issues related to menstruation, childbirth or sexual intercourse. This can be another potentially useful tool of nurturing. You may be able to find a clinic or spa that offers this treatment, but you can safely do it yourself at home as well.

After boiling up a pot of specifically chosen herbs, which could include sage (purifying), rose petals (feminine), chamomile (calming), hyssop (a cleansing herb mentioned multiple times in the Bible) or many others, you leave the herbs in the pot and place the pot under a seating situation where your vagina is exposed to the steam. Then you wrap a towel around your waist and adjust it so it traps the steam. The seating situation could be an ottoman with the cushion removed leaving a type of tube that you can sit on, a stool with a hole in it, a commode walker for elders or anything you can invent from the items already in your home. For those who wish to experiment with healing techniques privately, this is a more personal experience that can enable you to enjoy the same kind of nurturing to integrate the healing and purifying power of herbs to unblock congestion and restore healthy energy in the pelvic area. You can look upon this as a further step than taking long, hot baths, which also can feel soothing and nurturing. Vaginal herbal steaming adds a more therapeutic component to the experience.

As you make further choices on what you can do on your own to pursue deeper healing from sexual trauma, you don't always need to focus direct attention on your pelvic area to usher in a sense of nurturance and calm. Other approaches can also be effective.

Flower Essences

Flower essences, which are herbal infusions derived from the flowering part of plants, can also promote wellness for sexual trauma survivors. Flower essences, which were first formulated by Dr. Edward Bach in the 1930s, are energetic or vibrational remedies designed to imprint the life force of plants on those who take them in a way that promotes balance and harmony in the mental and emotional realms. Each plant evokes a specific quality, and you are empowered to use your intuitive sense to select the essences that you believe align with conditions or symptoms you may be dealing with and that may strengthen a particular quality that may feel weak or lacking in your being. Flower essences are a form of subtle vibrational medicine and when used correctly can shift the "vibe" that you give off, attracting better people and situations into your life.

Patricia Kaminsky, founder of FES Flowers, suggests that flower essences are best taken either just after you wake up or just before bedtime, because those are the times of the day when the spirit body is traveling to the spirit realm. You can search her website to find the correct flower essence for your symptoms. http://www.fesflowers.com

Other means to aid in your healing may be much simpler and easier to integrate, although again, you may not have yet linked such choices with the specific mission of healing from sexual trauma. Steering your diet toward healthier foods certainly can play a part in restoring a sense of well-being. When you are seeking to release and clear blockages and enhance the flow of energy, eating right just makes sense. I recommend a diet of cruelty-free, wholesome organic food, which can carry a greater nurturing quality if we take the time to bless it before we consume it. Every little choice in a positive direction matters.

At some point on your healing journey, you may want to consider going on a brief cleansing fast. It's certainly advisable to consult with a trusted medical practitioner and conduct your own research into the benefits and potential downsides to fasting as they relate to your body and life situation before deciding to go on a fast, to make sure you are in good physical health and aren't taking any medication or dealing with a condition such as an eating disorder or diabetes that would not match up well with a fast. But if fasting is determined to be something safe for you to do, you would be utilizing one more resource that is capable of raising your body's vibration. Fasting combined with flower essences is powerful vibrational therapy! Raising your frequency this way is another avenue of potential healing, and it can help you feel in charge of how you are approaching your path toward wholeness. It is said that dark energy or spiritual parasites cannot stay attached to the host body if it is vibrating at a high enough level. This is why the Bible often refers to various characters who fasted and prayed in order to connect with God. I would recommend fasting only after thorough research and in the later stages of healing.

Lifting the Dark Clouds

All throughout this book, we have been exploring many different approaches that can in some way contribute to lifting the dark clouds that so often consume our lives as sexual trauma survivors and clear the pathway for our divine spark to return. One action that you can take alone that can be especially helpful is something that is often recommended during therapy. The assignment is to choose a day that powerfully reminds you of your trauma, whether it was the specific date you were sexually assaulted or some other date that has taken on a negative meaning for you, and set out to rewrite the script of that day.

Imagine the biggest or most meaningful gift you could give to yourself and find some way to bring that into your life on that day. Maybe that's a trip you always wanted to take, a place you always hoped to visit or a significant accomplishment such as climbing a mountain. When you do that on the day that you associated with pain or suffering, and you give that dark day a memory with an even stronger positive emotional imprint, you are no longer owned by the negative force of that day. You've rewritten your script.

There's another gift that you might seek to give yourself. This one relates directly to that image of bringing a divine spark back into your life. Because soulful, divine love is the highest form of ways to grow the divine spark, ask for a spark of divine light from a donor. This could be someone who knows you well and unconditionally cares for your well-being. The seed or spark can come from the person of your choice through a warm, gentle hug, a gentle kiss or even a positive, loving gaze into your eyes. Sometimes when we are in pain, it is hard to look into the eyes of someone who is sharing with us real love and true compassion. It can often lead us to tears. But there is healing in that tearful release. You may be surprised

© Can Stock Photo / tanais

at people willing to do this for you, whether they happen to be a close friend, a loving family member, a therapist or a highly spiritual person who embraces an opportunity to cultivate the light for you. After you have received this spark, you can ground the experience in your being by practices such as meditation or Qigong that focus on cultivation of something wanted and welcome.

You can also ask someone who loves and cares about you to lend you or give you a sweater or blanket, so even when you are away from them, the caring person's energy covers you. Or it could be a symbol of strength that might serve you well, in which case a piece of jewelry or other item gifted from a person who symbolizes resilience could be a nice token. Humans have been imparting qualities of character to each other through symbolic transfers of energy via objects since the dawn of time. It's a small thing to do this for sure, but sometimes small things can affect our inner world in ways that create big shifts.

As you can see from our brief survey of other healing approaches that can serve to revitalize and reinvigorate survivors of sexual trauma, there really are a multitude of possibilities waiting for you. When you are ready to seek those that are right for you, a deeper and more profound healing will be awaiting you.

Cultivating Spiritual and Mental Health

As another dimension of nurturing your soul on your path towards healing, you may find yourself turning more closely to your spiritual beliefs and practices, or adopting some if you had none. There are many resources available—some based on time-tested traditions, and some written from a more contemporary perspective. Inspirational books offer wisdom that can help overcome obstacles and reorder thoughts in a positive way, even in the face of adversity, to facilitate the renewal of the human spirit. Some of these books are a "page-a-day" type and can offer a morsel of wisdom each day to inspire us. There is a saying in Oriental Medicine "Qi flows where attention goes." If the mind is stuck in a negative loop or a downward spiral, it can certainly hinder healing. Studying spiritual wisdom that has stood the test of time, that our ancestors have recorded for us, allows us to avoid many common pitfalls. Every small step in a positive direction matters as we create our brighter future. Rather than treating mental illness, I've always found it preferable to cultivate mental health, and time-tested spiritual wisdom, whether it is found in a book or in the words of a favorite speaker you find online, can nourish the soul and lead you to a better place.

May you walk in your journey in Love and Grace.

Chapter 6
Reclaiming Your Radiant Female Essence

At the start of this book, I shared with you my desire that whoever you are, and no matter when you or someone you care about may have suffered sexual trauma, you would come away from our time together firmly grasping ideas and inspiration regarding new possible avenues for you to travel down in the pursuit of your physical, mental, emotional and spiritual healing. If you were feeling stuck in familiar patterns of suffering, or hopeless about your future, I trust that you now recognize the many viable options you can choose from to get confidently moving on your journey toward health and wholeness. In the previous chapter, we touched upon a wide spectrum of body-oriented therapies that can be effective in treating the effects of sexual assault, to build on our exploration into the powerful potential of Oriental Medicine to treat sexual trauma. Your map for healing, hopefully, truly has become much bigger, wider, full of tools and resources to call upon.

Still, you may have noted one especially relevant domain of healing from sexual trauma that has not yet been given our full attention. It's time to shine the light on the multi-dimensional process in which survivors can reclaim their full, radiant female essence. We're going to zero in on the potential for moving from sexual trauma to sexual healing. And we'll even dare to imagine a world in which women and men may begin to come together as loving partners and allies in caring for the needs of sexual trauma survivors and growing in awareness of the sacred possibilities of a healthy sexual union.

The Blooming of the Flower

Earlier, we used the analogy of the petals of a flower being torn or crushed as one of many ways to understand what happens to women when they are sexually assaulted. Now the question to consider is this: can this "flower" (the healthy unfolding of your sacred sexual energy) ever bloom again?

I want to assure you that you really do have the potential to resurrect the healthy unfolding of your sexual energy once more. You can regenerate what was lost and reclaim your sexual essence. You can blossom again and shine in your full strength and natural beauty when the time is right.

I believe that if a woman can be compared to a flower or any plant, it can be said that she is naturally more perennial than annual. She has the capacity to exhibit that radiance and beauty through multiple cycles. The good news for us as survivors is that everything we do to promote our healing, working with the tools and resources that we have discussed and any other effective approaches we may discover that address your pain and loss, can become the key that unlocks the gate to a new cycle of strength and radiance.

Now, that doesn't mean that the experience of sexual assault left only a minimal impact. To understand what happened to that cycle of blooming and regeneration when you were sexually assaulted, we are going to stick with this flower analogy a bit longer.

We know that a real flower blooms and has a life span. If you want a flower to remain as beautiful as possible, the best thing to do is to leave it rooted to the ground so it may have the earth, sun and rain to draw upon. If you cut a flower and take it into your home, it still has a lifespan but it will not be as long.

We as human beings are also rooted to the ground through energy points on the soles of the feet, and also connected to higher realms above through an energy point on the crown of the head. But when we suffer sexual trauma, those connections can be broken—until we find the way to repair them. Something so natural, something potentially so wondrous, was disrupted by the violation that occurred, a violation that is so antithetical to how things are supposed to be.

You can look at human sexuality through the lens of a commonality it shares with what happens with bees and flowers. Bees typically fly from flower to flower, and every once in a while they land on a flower that truly captures their attention. They decide that this is the flower that's right for them—there's something about the fragrance, the color, the climate it chooses to be in that just pulls them in. You can say that it's at least somewhat similar to men showing interest in many women until finding that "flower" that especially attracts them.

Opening When She's Ready

On the female's part, you may recall our earlier discussion about how the etheric template of the heart is actually what's broken when we say that someone "broke her heart." Well, there is also an etheric template around a woman's uterus, her female sexuality. The energy flow is ignited via the consciousness, travels down the throat, traverses the heart, goes through the whole body and then opens in the pelvic area, almost as if a subtle energy trumpet shaped flower lives within her.

Now, in a healthy situation in which a female is fully and energetically consenting to sexual contact, the flower opens when it is ready. Then her Yin Jing sexual essence, much like the nectar of a flower, starts to activate her sexual organs.

Men respond to this process when they tune in to her pheromone signals, and it is natural that they would. In the male-female sexual exchange, he brings his yang energy, which ideally carries the warmth of the sunshine and is bathed in love. This is what encourages the sexual unfolding, and as a result he obtains her Yin essence,

which nourishes his Yang and provides him greater energy to go out in the world and be all of who he is. The female gains from the deliverance of that Yang energy, which energizes her and brings her joy and furthers a cycle in which she can produce more and more of that special Yin female essence. In a healthy, loving relationship, she is capable of producing enough Yin sexual essence that it can last throughout her adulthood, which is what happens in a healthy and loving marriage.

In this picture of balance and harmony, the male is the pollinator who, with love and reverence, has approached the flower with his pure yang energy via rhythmic movements in a variety of tempo and amplitude, much like a bow interacting with a violin. He has brought his seed with the right kind of energy to produce fruit and enable her to experience multiple flowerings where she shines more of her radiance into the world.

However, in a sexual assault, where she never willfully consented to the act, her etheric template protecting her yin female essence was never let down. The perpetrator forcibly broke through. From the perspective of Oriental Medicine, you could say that this forceful act damaged the flow of energy in the meridians, which in this image would be the tiny veins in the plant that nourish the leaves and petals of the flower. It's no wonder, then, that the flower's potential lifespan was blunted by the assault. Instead of the perennial's natural multiple flowerings, the damaged flower was left with the temporary inability to function properly and to produce radiance and beauty.

That severe damage, that crushing of our flower, is one major reason why the challenge of healing for survivors can at times seem overwhelming. But all through the journey toward wholeness, simply believing that reclaiming your radiant female essence is possible, and that many other women before you have done this, can strengthen and fuel your healing process. As you find that a wound that once seemed lasting and permanent can indeed be treated, you really do begin to rediscover your radiance as a woman.

Is It Safe to Shine Again?

The question that still may hover around your heart and your being is this: after once having your sexual system crushed and torn, will I ever decide that it is safe to open to another man? Can I really share my radiance with someone else in that intimate way?

That is a question for which there may be no answers on the treatment table of an acupuncturist or any other body-oriented healer. The discovery of those answers for yourself as one survivor, and for our collective state as women, perhaps can begin through an exploration into our cultural mindsets and influences and what we can begin to do to bring positive change where it may be so badly needed.

Maybe you have already been blessed with finding a loving partner who understands the blossoming process and has co-created a space with you in which you feel you can fully share your female essence. My hope for you is that you may continue to shine your energetic beauty in the way that fulfills your soul. You also may find that you can reach out to and encourage other survivors of sexual trauma on the healing trail. Conversely, if this is not a picture that has emerged for you as a survivor, tune in now as we examine the issues that you may or may not have considered.

Like it or not, we live in a society in which there are powerful cultural forces and programming that can make it difficult for men to approach the flowering of female essence in the right manner and spirit. For example, look at the language often associated with the male need to succeed or conquer in sports, business or many other life domains. These words are "destroy" or "own." We can also look at common slang used around sexuality, words like "pound," "drill," "hammer," "nail" or

"bang." These all come from carpentry or construction. Not unusual when for millennia, the male had the voice to define experiences and the female voice was suppressed. When the female voice finally started to rise, the cry "sexual abuse" and "rape" was common. All too often, the male who operates according to this old paradigm programming carries that kind of mindset into preying on females and seeks to engage in a sexual union with them using methods that involve coercion, overpowering and deception. In the act of approaching the female flower in hopes of obtaining the nectar or essence inside, destroying, drilling or hammering hardly seems conducive to the female opening. The unfolding flower of female sexuality just doesn't fare well when pounded, banged or screwed. The female sexual system is designed to unfold in the presence of love. On the shadow side, yes, we all have a shadow, it unfolds for money or material gain as in the case of prostitution or gold digging, which may be fueled by either survival or materialism.

Some men can get caught up in the destructive belief that successfully hunting many women and treating the sexual act this way somehow proves their male strength. Well, if I were looking at flowers in my garden and said, "Oh look, it's a flower. I can crush it because it is weak," would I be proving that I'm strong? No, I would be proving that I'm a brute.

Does that mean that all men are potential perpetrators, men whose goals cascade all the way down to the realm of wanting to terrorize and humiliate? Absolutely not! There are many men out there who resist or rise above cultural programming to understand the need to "play the bow of the violin" with women to whom they are attracted. I salute them. Unfortunately, it often seems to survivors looking out at the cultural landscape that vast numbers of men around them still appear guided by those aggressive tendencies that distort the pure yang energy.

In their zest to succeed in every endeavor, many men, especially those who are younger, see women as targets to be preyed upon and conquered. Rather than join in loving harmony in an act that balances the yin and yang, they are propelled by that conquest mentality. How often do we hear stories of teens and younger adult men who engage in contests to see who can "deflower" the most girls and young women? They are the bees who are not at all inclined to land on only that one flower that draws them in but who rather buzz and buzz around constantly, stopping at any one flower only long enough to do what they feel they need to do and quickly move on to the next target. In other words, they are driven by the beauty and sex that multiple women provide, without the burden of an ongoing commitment to fully exchange the essence of what the male and female both bring to the picture.

In some circles, a harmful cycle sometimes is born from this destructive mindset. Women who have been used in this kind of union begin to lose some of their natural radiance. Meanwhile, the men who only want to drain female sexual essence go on to prey on the next female that still appears to carry that full, natural radiance and beauty. The women they leave behind become characterized by other preying men as "damaged goods." These women live with varying degrees of trauma or betrayal, even if the sexual act that was shared did not rise to the level of sexual assault.

Tuning to the Frequency of Love

Again, that's an extreme scenario. But, as we move on from the flower analogy, I believe it is fair to say that a woman's sexual experiences, especially when she is younger, leave a significant imprint on her being. To turn to another analogy that I introduced briefly in an earlier discussion, the human body is like an instrument, perhaps a guitar. From the first time you enter into a sexual experience, you're tuning that guitar to play music in a certain way from that point

on. Naturally we all want to tune the guitar to the frequency of love. Women who experience real love as part of their earliest sexual experiences are going to have a much easier time hitting the high notes in life. Those positive exchanges, guided by mutual respect and caring, set an emotional tone for their future. They are imprinting their soul and encouraging their divine spark to flourish.

Women who have an early negative experience, which may even come in the form of rape or child sexual abuse, are going to be tuned to a low frequency and will consequently face significant challenges in seeking to hit those high notes in life. It will take a greater effort to turn the loss to a positive direction.

I watched a movie that illustrated this dynamic perfectly. In Girl with a Pearl Earring, Scarlett Johansson plays the role of Griet, a young servant in the household of a painter in 17th century Holland. The painter's rich patron notices the beautiful young woman and immediately decides that he wants to have her. Since she is not willing and she is repelled by him, he intends to rape her.

She understands the threat and makes a fascinating decision. The son of a butcher in her town has shown his own interest in her, and she can sense his sincerity and caring. Without waiting for any romantic foundation, she chooses to have her first sexual union with him. I totally understood and appreciated her decision. She knew that the prospects for her life going forward would be much better if her guitar was tuned by this sincere and loving young man rather than the predator closing in on her.

Transforming the Gender War

As survivors of sexual trauma, we know that there are far too many predators out there. They assaulted us, tuning our guitar, our body instrument, to that lower frequency. Through our healing, we are doing everything in our power to reverse the effects of that low-frequency tuning. We are using healing techniques to rewrite the histories of our bodies. And we certainly don't want to feel we will be faced with the need to fend off other potential perpetrators in the future. Ideally, we'd like to believe that there are men who can understand and support us on our journey, to assist us in our healing from the past. Looking to our future, we would also like to carry hope to someday find a loving partner fully capable of entering into a sexual union and a partnership in the true spirit of love. All survivors have their own unique goals, beliefs and approaches to sexuality, but for many survivors of sexual trauma, entering into such a loving partnership at some future point would represent a final touch, a pinnacle in their healing journey.

Can this happen during a period when it often seems that we're hopelessly wrapped in an endless gender war, with rape and sexual assault one terrible component? Other teachers and guides who have far more expertise than I do have written many books on this subject, but from my lens as a survivor and a woman committed to doing whatever I can do to encourage and inspire hope and healing for other survivors, I will offer a few thoughts.

For one thing, de-escalating the gender war is a challenging goal that will take a gradual changing of old paradigms. Doing so will mean seeking to open a new door into different ways of thinking, understanding and relating. As female survivors of sexual trauma, we of course deserve and hope for and envision a better understanding from men. We also can help our own cause by offering empathy and understanding for the ingrained attitudes and beliefs that often burden men.

Now, before I go any further, let me be very clear that I am in no way arguing that women who have been raped or sexually assaulted by men must forgive their perpetrators. That person or persons who violated you, no matter what the circumstances of the assault may have been, have in most cases not earned your forgiveness and never may. Forgiveness requires atonement and restitution—a sincere change of heart whereby the perpetrator simply cannot repeat such a deed again, along with compensation to ease the life of the person who was taken from. Modern new age forgiveness often forgets about atonement and restitution and instead pressures the victim to forgive or, even worse, to verbally express forgiveness to someone who is unrepentant. In fact, I sometimes wonder if the mandate to forgive, which unfairly burdens survivors of incest, rape and other abuses was invented by those that wanted to "trespass against us." That said, I am highly supportive of ministries that help to heal the souls of sex offenders, but I also believe that some offenders are beyond repair. For some few, correct pastoral guidance can reorder the thinking and behavior patterns of sex offenders and help them regain a new life that will lead to true fulfillment.

If your own journey has brought you to a place where you are able to cleanly and fully tap into an energy of forgiveness, and this is something that feels right in your soul, then it was the correct spiritual remedy in your unique case. But for many survivors, forgiveness in this situation is something that our whole being informs us, subtly or via a screaming voice, just does not fit. Sometimes we may feel pressured to forgive our perpetrator by other people or through beliefs from some religious or spiritual tradition we follow. My guidance is to follow the path that best makes your spirit feel strong and whole.

Trying to force forgiveness early on, when that's not something you feel in your heart, could energetically invite new layers of trauma. When you've worked so hard in your commitment to your deep healing, you certainly don't need that. It can be enough to just accept that something so wrong happened to you, that you can't deny its full impact, and that you are taking steps to set upon a new course that will align with your physical, emotional and spiritual needs. Forgiveness does not need to be added to the equation. If it does come up in your soul organically, it will likely be because you are able to view the event after enough time has passed, and you have healed, from a place of much higher consciousness, taking into consideration many more cultural dynamics that brought about the negative event.

Now, returning to the idea of looking at men through more of a collective lens, at some point along the path of your healing you may find that you are no longer burdened by those dark voices that say "All men are dangerous," "I can never trust men again" or "Men are not safe to enter into any kind of sexual union." Perhaps you will find room in your being to look around the picture of our gender war and wonder what it's like on the "other side."

Understanding Male Trauma

We recognize that men descend from generations of hunters who in days past had to hone those hunting instincts and kill prey to feed their families. The practice of hunting over millennia developed in men a keen predatory eye. Even today, some men are engaged in what seems to me like a barbaric practice of how animals are killed and slaughtered for producing meat to consume. If the hunting of animals is pursued in the Native American tradition of only killing for need and taking the time to honor the spirit of the animal and express gratitude for its sacrifice before the act to end its life, it will not lead to energetic wounding of the person who killed. But that's not the way that animals are killed in the modern meat industry in our culture.

There are many other domains in which men may feel pressured into a state of wanting to destroy and overpower. Many men, and more recently, some women, are still trained to kill in war and other acts of aggression in the name of their country and people. As we mentioned earlier, men are still oriented toward going out and trampling their opposition in sports and business. It's far too common for men to carry a belief that they have only succeeded in life as a male if they have somehow or other found and adhered to a way to come out on top, that they have successfully beaten or conquered whoever or whatever stood in their way on their path toward achievement and fulfillment. Of course, such dynamics have a positive as well as a negative side. However, when I think of the impact on a man's soul to live with the burdens of these pressures, I am able understand their perspective.

Then there's the whole matter of circumcision. When I look around, I see growing evidence that circumcision is becoming a hotter topic for discussion and debate these days. I find this encouraging because I believe this is something that needs to be talked about. If you've never seen what the standard circumcision procedure looks like, it may be time to learn. Usually the baby boy is strapped down on a board with some form of restraint. Sometimes the baby is given an anesthetic, but usually not. There's usually a great deal of crying and screeching from that baby who is experiencing a tremendous amount of pain in the genital region for having part of his penis cut off. He also is experiencing the deeper energetic wound by the decision of his parents or caregivers to subject him to something so contrary to a baby's needs in the early development cycle, when he is completely dependent on the nurturing protection of his mother for a healthy outcome as a child and later as an adult.

Those who advocate for circumcision insist that it's a necessary procedure for hygiene. I find it hard to accept that argument because if we can wipe our bottoms with toilet paper, a man can certainly manage cleaning his penis, which happens to be in the front. I also understand that there may be religious beliefs that influence parents' decisions regarding circumcision. However, it's staggering to realize that a vast majority of baby boys are still being circumcised today.

I am heartened when I read or hear about change agents seeking to help men and women understand that circumcision causes significant trauma for a boy, and it's a trauma that he may carry fully into adulthood. The trauma has been driven deep into the baby's unconscious, and I can't help wondering if it could contribute to reduced sensitivity and even an aggressive manifestation of male sexuality. And this influence is not something males had any choice over. It was something that was simply done to them according to the direction of others.

Some males will not identify any negative impacts from circumcision with respect to their well-being as an adult, but for others, seeing circumcision for what it is, namely male genital mutilation, may explain some harmful ingrained attitudes and behaviors.

So that's another factor in the influences that can shape men's lives. Many men today also simply struggle with excessive yang energy, which can translate into that desire to buzz from one sexual encounter to another. Hopefully, more men can discover that a healthier and more natural way to burn off that excess yang is to spend more time in the gym or to find work or hobbies in which they employ their hands and use a lot of physical energy. They could also benefit from eliminating or decreasing alcohol use and eating a healthy diet, with much less red meat consumption because it can contribute to the yang being out of balance.

Again, I want to emphasize that none of these influences or areas of cultural programming in any way, shape or form offer an excuse for men to ever act as a perpetrator against another human being, but they could offer a partial insight into the causes. Some professionals who deal with the effects of sexual trauma suggest that many perpetrators were sexually traumatized themselves and carry depths of anger, shame, bitterness and hatred. That may be true, but it's also true that they chose to act as they did in a way that damaged others rather than pursue and commit to their own healing.

Our perspective regarding all those other men out there who would never rape or abuse someone, but who do carry the effects of childhood wounds or toxic programming, can be geared toward support, encouragement, understanding, empathy or just hope for change. This kind of outlook would not only benefit men but ourselves as women too. The way I see it, the circumstances of women in our culture will not change until the circumstances of men change as well. We are interdependent and interconnected. You simply can't liberate women without also liberating men. If the liberation of women from gender-based suffering causes undue suffering for males, it will never work. It has to be a dance that liberates women and at the same time brings men to a better, higher, freer place.

That means that it's in our interests to contribute in any way in which we think we can help and feel an openness in our heart to do so, even if it's just rooting from the sidelines in any discussion or endeavors that can help stir an evolution for the genders that is long overdue. Sometimes I wonder whether we might begin to see noticeable change in the behavior of males in our culture if every man re-established a healthy relationship with the land and directed their powerful Yang energies into the Earth so our planet can continue to nourish us all. I am quite sure that the Swords into Ploughshares statue (Inspired by the Bible quote:

81

Isaiah 2:3–4 [1]) outside of the United Nations is suggesting the same thing.

Dmytro Klymchuk - Ukraine © 123RF.com

---

1 And many people shall go and say, Come ye, and let us go up to the mountain of the Lord, to the house of the God of Jacob; and He will teach us of His ways, and we will walk in His paths: for out of Zion shall go forth the law, and the word of the Lord from Jerusalem. And He shall judge among the nations, and shall rebuke many people: and they shall beat their swords into ploughshares, and their spears into pruning hooks: nation shall not lift up sword against nation, neither shall they learn war any more.—Isaiah 2:3–4

So What is Going Wrong?

The dominator mindset, which is based in sexual entitlement and disregard for the female's experience, and disregard for her consensual participation in the sexual act, creates trauma in females. A healthier scenario involves patience, consent and the earning of trust whereby the female will take down her boundaries gradually in a healthy manner and become ready to receive the male. The first example of the dominator mindset actually creates so much trauma and disruption in the female's sexual system that it becomes difficult for her to regenerate sexual energy. In colloquial language, the male will describe the female as being "used up" or "damaged goods." However, in many cases it is the male, because of his ignorance or aggression, that caused the damage. If a woman is not traumatized, she is able to generate much more sexual energy throughout her lifetime than if she has to intermittently sequester herself for healing. As such, it behooves males who don't understand to at least learn about how a normal female's sexual system works, and see that, by not traumatizing females, they are acting in their own self interest. But many remain unteachable, externalizing problems, refusing to look inward, insisting rather that they can just prey on more females or children to meet their needs.

A wise teacher of mine observed that in males, the body's lower energy centers become activated first when in the presence of a female they find sexually exciting. In females, her upper energy centers need to be activated before her sexual system becomes active. Males and females are like opposite sides of a magnet. For this reason, if a female is looking for loving sexuality as opposed to sex without love, it is important that she withhold sex until the male's heart center has been activated. A getting to know each other period, over time where each interaction leaves both parties feeling good, will often and eventually result in a heart connection forming, after which loving sexuality can be shared. As a trauma survivor, it is wise to be extra cautious when getting into your first post-trauma relationship. It is a fact of how nature works that predators can sense the presence of wounded people just as it happens with animals in the wild. This is why I strongly suggest to pursue a sincere multi-leveled healing process and proceed cautiously when re-entering the dating world. It is wise to be patient and take the time to get to know your potential partner, and even invite the individual to meet family or friends first and let your family or friends screen the individual for potential red flags. A predator will most likely turn down an invitation to meet friends or family, which will save you from wasting energy with such a person. It is also wise to consider the potential partner's family of origin, and relationships at work and with peers. Do not be hasty. Real love is patient and should take time to blossom.

A Hand Can Hurt or a Hand Can Heal

There are men who commit to a deeper understanding and support of women and everything that they bring to the table. These men are compassionate and empathetic in a way that enables them to truly listen to us and assist us in our healing journey. How wonderful it would be for all of us if more and more men came to recognize that while the hand can be a fist that punches, it can also be a hand that heals. Just as a man is able to "cleave" harmful spiritual energy into a woman's body, a man of good character, who lives a wholesome life and engages sexually with a woman in the vibration of love, can equally cleave positive, uplifting spiritual energy into a woman's body. Conscious, patient, loving sexuality heals!

In sexuality, men do have the capacity to embrace an approach in which they are the bow playing a violin.

They can commit to helping a woman tune to a higher frequency, to encourage her regeneration and the reclaiming of her female sexual essence. In this spirit, the sexual act can be a healing and uplifting experience, understood to bring wave-like energetics, like the ebb and flow of the tide. A "crashing" wave is orgasm.

Of course, grasping and following such approaches to sexuality may not come easily or naturally for those who have watched hours and hours of porn. I wince in pain whenever I consider how the men in many porn films appear to celebrate the breaking of the woman's spirit. This is reminiscent of the old-school barbaric "breaking the horse." In fact, the language of prostitution in which a brothel is referred to as a "stable of prostitutes" or the many brothels in Nevada with "ranch" in the name echoes the horsemanship and prostitution connection. How can anyone think that breaking someone's spirit is a natural part of human sexuality? Yet we live in a culture in which millions of men, and many women too, for better or worse, take their sexual cues from porn. Healing needs to happen not only for individuals, but for society.

Fortunately, there are other choices for sources of what constitutes a real and satisfying approach to sexual union. Ancient practices such as Tantra and Taoism teach us that sex is a natural, vital energy, a creative force that we can tap to heal and regenerate our body's energy while connecting to the greater universal energy outside of ourselves. Through a loving union with another, we can harness and cultivate more energy, not just in the sexual experience but in our everyday physical, mental and spiritual life.

As women and men, we may not realize our true potential in the wondrous realm of sexuality until we evolve to a higher level. It's a goal worth holding in our personal and collective vision.

The Golden Thread of Hope

Survivors of sexual trauma can call upon other helpful resources geared specifically toward their needs and challenges in bringing their full selves to a sexual union. Wendy Maltz's book The Sexual Healing Journey assures women who have been raped or who survived childhood sexual abuse that sex "is a natural biological drive, powerful healing energy, a vital life force" and that it can be "respectful, honest, mutual, intimate, safe, empowering, a choice."

Again, there is so much out there available for survivors who are sincerely interested in traveling a path toward a wholesome, healthy, nurturing and joyful sexual relationship. I know that the forces of doubt or cynicism can seem overpowering because we are all living through a difficult period in our culture and world. At times it can feel like we're heading for a collective collapse, and the only way to move toward a new paradigm is to start building it up from the ashes. Fortunately, there are those among us who even in our dark times have the courage and the wisdom to carry the golden thread of hope. We can make the choice to grab hold of that thread.

In my life today, I take the time to volunteer at my local animal shelter and I have rehabilitated several "chain dogs." It's an absolute joy every time I witness the inspiring truth that any dog, no matter how much or how long it has been abused or mistreated, has the capacity to heal if given the right care, nourishment and environment. To wag his tail, to run, to play, to love and be loved again. Because I've seen and facilitated this transformation in animals, I know very well that it is possible for humans.

There is indeed hope for all of us, too, to thrive again, to joyfully find our place in the flow of life, to love others and to be loved by others. That's a sacred truth that you now have learned more deeply. May you now go out and live that truth!

© Kris Wiltse Illustration

HEALING SEXUALITY WITH HER BELOVED

While the masculine fire can be an inferno that scorches, it can also be the warmth of the sun which provides nourishment for the female sexual flower.

© Kris Wiltse Illustration

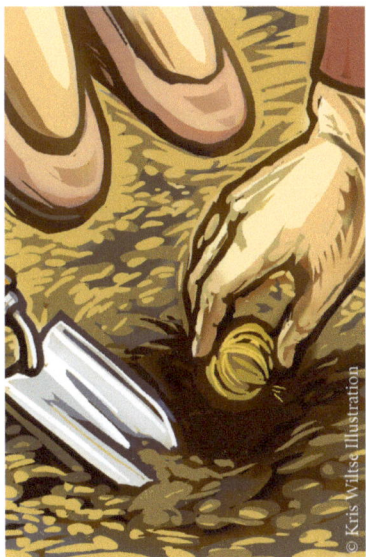

FARMER AND LILY PLANT TOGETHER

Real affection requires time, trust, patience and nurturing before it can blossom into love.
Much like the flower bulb that Lily and the Farmer plant and water together,
the right conditions are needed for the flower to bloom.

Contact the Author

*I want to thank you from my heart, for investing your time and attention
to consider what I wish to offer to the cause of healing sexual trauma.
If you choose to work with an acupuncturist in your geographic area,
I am may be available to assist through consultations, where desired and appropriate.*

*I'm also always open to hearing from other professional caregivers,
whether you happen to be a practitioner in Oriental medicine, or another
body-oriented modality, or you play another role in any system
where the care of survivors is important to you.*

*Krisztina Samu, LAc
info@projectacuhope.com*

Bibliography

Arvigo, Rosita and Epstein, Nadine. *Rainforest Home Remedies: The Maya Way to Heal Your Body and Replenish Your Soul.* New York: Harper One, 2001.

Baker, Elsworth, M.D. *Man in the Trap - The Causes of Blocked Sexual Energy.* New York: The Macmillan Company, 1967.

Baldwin, William J. *Spirit Releasement Therapy: A Technique Manual.* Terra Alta, WV: Headline Books, 1992.

Brennan, Barbara. *A Guide to Healing Through the Human Energy Field.* New York: Bantam, 1988 and *Light Emerging: The Journey of Personal Healing.* New York: Bantam, 1993.

Chia, Mantak and Chia, Maneewan. *Healing Love through the Tao: Cultivating Female Sexual Energy.* Rochester, VT: Destiny Books, 2005.

Chia, Mantak. Karsai Nei Tsang: *Therapeutic Massage for the Sexual Organs.* Rochester, VT: Destiny Books, 2011.

Farrell, Yvonne. *Psycho-Emotional Pain and the Eight Extraordinary Vessels* Singing Dragon; 1 edition, 2016

Greenwood, MD (MB), Michael. *Possession:* Medical Acupuncture; Vol 20, Number 1, 2008:

Jarrett, Lonny. *The Clinical Practice of Chinese Medicine.* Stockbridge, MA: Spirit Path Press, 2006.

Kaptchuk, Ted. *The Web That Has No Weaver: Understanding Chinese Medicine.* New York: Congdon & Weed, 1983.

Shapiro, Francine. *Eye Movement Desensitization and Reprocessing (EMDR) Therapy, Third Edition: Basic Principles, Protocols, and Procedures.* The Guilford Press 2018

Van der Kolk M.D., Bessel. *The Body Keeps the Score: Brain, Mind, and Body in the Healing of Trauma:* Penguin Books; Reprint edition (September 8, 2015)

Ogden, Pat; Minton, Kekuni; Pain, Clare. *Trauma and the Body: A Sensorimotor Approach to Psychotherapy (Norton Series on Interpersonal Neurobiology)* W. W. Norton & Company; 1 edition (October 13, 2006)

© Can Stock Photo / tanais

www.ingramcontent.com/pod-product-compliance
Lightning Source LLC
Chambersburg PA
CBHW041449210326
41599CB00004B/189

9 780578 214818